Applied and Numerical Harmonic Analysis

Applied and Numerical Harmonic Analysis

J.M. Cooper: *Introduction to Partial Differential Equations with MATLAB* (ISBN 0-8176-3967-5)

C.E. D'Attellis and E.M. Fernández-Berdaguer: *Wavelet Theory and Harmonic Analysis in Applied Sciences* (ISBN 0-8176-3953-5)

H.G. Feichtinger and T. Strohmer: *Gabor Analysis and Algorithms* (ISBN 0-8176-3959-4)

T.M. Peters, J.H.T. Bates, G.B. Pike, P. Munger, and J.C. Williams: *Fourier Transforms and Biomedical Engineering* (ISBN 0-8176-3941-1)

A.I. Saichev and W.A. Woyczyński: *Distributions in the Physical and Engineering Sciences* (ISBN 0-8176-3924-1)

R. Tolimierei and M. An: *Time-Frequency Representations* (ISBN 0-8176-3918-7)

G.T. Herman: *Geometry of Digital Spaces* (ISBN 0-8176-3897-0)

A. Procházka, J. Uhlíř, P.J. W. Rayner, and N.G. Kingsbury: *Signal Analysis and Prediction* (ISBN 0-8176-4042-8)

J. Ramanathan: *Methods of Applied Fourier Analysis* (ISBN 0-8176-3963-2)

A. Teolis: *Computational Signal Processing with Wavelets* (ISBN 0-8176-3909-8)

W.O. Bray and Č.V. Stanojević: *Analysis of Divergence* (ISBN 0-8176-4058-4)

G.T. Herman and A. Kuba: *Discrete Tomography* (ISBN 0-8176-4101-7)

J.J. Benedetto and P.J.S.G. Ferreira: *Modern Sampling Theory* (ISBN 0-8176-4023-1)

A. Abbate, C.M. DeCusatis, and P.K. Das: *Wavelets and Subbands* (ISBN 0-8176-4136-X)

L. Debnath: *Wavelet Transforms and Time-Frequency Signal Analysis* (ISBN 0-8176-4104-1)

K. Gröchenig: *Foundations of Time-Frequency Analysis* (ISBN 0-8176-4022-3)

D.F. Walnut: *An Introduction to Wavelet Analysis* (ISBN 0-8176-3962-4)

O. Bratelli and P. Jorgensen: *Wavelets through a Looking Glass* (ISBN 0-8176-4280-3)

H. Feichtinger and T. Strohmer: *Advances in Gabor Analysis* (ISBN 0-8176-4239-0)

O. Christensen: *An Introduction to Frames and Riesz Bases* (ISBN 0-8176-4295-1)

L. Debnath: *Wavelets and Signal Processing* (ISBN 0-8176-4235-8)

J. Davis: *Methods of Applied Mathematics with a MATLAB Overview* (ISBN 0-8176-4331-1)

G. Bi and Y. Zeng: *Transforms and Fast Algorithms for Signal Analysis and Representations* (ISBN 0-8176-4279-X)

J.J. Benedetto and A. Zayed: *Sampling, Wavelets, and Tomography* (0-8176-4304-4)

E. Prestini: *The Evolution of Applied Harmonic Analysis* (0-8176-4125-4)

O. Christensen and K.L. Christensen: *Approximation Theory* (0-8176-3600-5)

Ole Christensen
Khadija L. Christensen

Approximation Theory

From Taylor Polynomials to Wavelets

Birkhäuser
Boston • Basel • Berlin

Ole Christensen
Technical University of Denmark
Department of Mathematics
2800 Lyngby
Denmark

Khadija L. Christensen
Technical University of Denmark
Department of Mathematics
2800 Lyngby
Denmark

AMS Subject Classifications: 40-01, 41-01, 42-01, 42C40
Cover design by Alex Gerasev, Cambridge, MA.

Library of Congress Cataloging-in-Publication Data
Christensen, Ole, 1966-
Approximation theory : from Taylor polynomials to wavelets / Ole Christensen, Khadija Laghrida Christensen.
p. cm. – (Applied and numerical harmonic analysis)
Includes bibliographical references and index.
ISBN 0-8176-3600-5 (acid-free paper)
1. Approximation theory. I. Christensen, Khadija Laghrida, 1963- II. Title. III. Series.

QA221.C495 2004
511'.4–dc22

2004043741
CIP

ISBN 0-8176-3600-5 Printed on acid-free paper.

Birkhäuser 

Printed in the United States of America. (SB)

9 8 7 6 5 4 3 2 SPIN 11382317

www.birkhauser.com

To Jakob and Sara

Contents

Preface **ix**

1 Approximation with Polynomials **1**
1.1 Approximation of a function on an interval 2
1.2 Weierstrass' theorem 4
1.3 Taylor's theorem 6
1.4 Exercises 13

2 Infinite Series **15**
2.1 Infinite series of numbers 16
2.2 Estimating the sum of an infinite series 21
2.3 Geometric series 24
2.4 Power series 27
2.5 General infinite sums of functions 34
2.6 Uniform convergence 40
2.7 Signal transmission 44
2.8 Exercises 48

3 Fourier Analysis **51**
3.1 Fourier series 52
3.2 Fourier's theorem and approximation 57
3.3 Fourier series and signal analysis 62
3.4 Fourier series and Hilbert spaces 64
3.5 Fourier series in complex form 67

3.6 Parseval's theorem 69
3.7 Regularity and decay of the Fourier coefficients 71
3.8 Best N-term approximation 73
3.9 The Fourier transform 76
3.10 Exercises 80

4 Wavelets and Applications **83**
4.1 About wavelet systems 84
4.2 Wavelets and signal processing 92
4.3 Wavelets and fingerprints 98
4.4 Wavelet packets 103
4.5 Alternatives to wavelets: Gabor systems 103
4.6 Exercises 104

5 Wavelets and their Mathematical Properties **105**
5.1 Wavelets and $L^2(\mathbb{R})$ 106
5.2 Multiresolution analysis 107
5.3 The role of the Fourier transform 107
5.4 The Haar wavelet 109
5.5 The role of compact support 122
5.6 Wavelets and singularities 123
5.7 Best N-term approximation 126
5.8 Frames 128
5.9 Gabor systems 132
5.10 Exercises 136

Appendix A **137**
A.1 Definitions and notation 137
A.2 Proof of Weierstrass' theorem 138
A.3 Proof of Taylor's theorem 141
A.4 Infinite series 145
A.5 Proof of Theorem 3.7.2 148

Appendix B **149**
B.1 Power series 149
B.2 Fourier series for 2π-periodic functions 150

List of Symbols **151**

References **153**

Index **155**

Preface

This book gives an elementary introduction to a classical area of mathematics — approximation theory — in a way that naturally leads to the modern field of wavelets. The main thread throughout the book is the idea of approximating "complicated expressions" with "simpler expressions," and how this plays a decisive role in many areas of modern mathematics and its applications.

One of the main goals of the presentation is to make it clear to the reader that mathematics is a subject in a state of continuous evolution. This fact is usually difficult to explain to students at or near their second year of university. Often, teachers do not have adequate elementary material to give to students as motivation and encouragement for their further studies. The present book will be of use in this context because the exposition demonstrates the dynamic nature of mathematics and how classical disciplines influence many areas of modern mathematics and applications. The book may lead readers toward more advanced literature, such as the other publications in the *Applied and Numerical Harmonic Analysis* series (ANHA), by introducing ideas presented in several of those books in an elementary context.

The focus here is on ideas rather than on technical details, and the book is not primarily meant to be a textbook. However, it may be used as such in courses on infinite series and Fourier series, where ideas and motivation are more important than proofs. In such courses, it will be natural to use some of the material from the two wavelet chapters as a guide toward more recent research, and as inspiration for further reading. The purpose of these two chapters is to present wavelets as a natural continuation of the material from the previous chapters. We are not aiming for a comprehensive treatment, and for these reasons, fewer exercises accompany the wavelet chapters as compared to the number of exercises included after the first three chapters.

We have attempted to organize the book in a way that makes the information accessible to readers at several levels. Basic material is therefore placed at the beginning of each chapter leading the reader toward more advanced concepts and topics in the latter parts of each chapter. Only selected results are proved, while more technical proofs are included in an appendix.

The first chapter, on approximation by polynomials, is very elementary and can even be understood by advanced high school students (however, the proofs may require more mathematical background). Furthermore, this introductory chapter gives the reader an idea about the content of the rest of the book. The next chapter, on infinite series, deviates from most other introductions with its strong emphasis on the approximation-theoretic aspect of the subject. It also supplements the quite rudimentary treatment most introductory university courses give to the subject nowadays: it contains several (classical) entertaining examples and constructions, which are rarely mentioned in class.

Wavelet analysis can be considered as a modern supplement to classical Fourier analysis, and for this reason we give a more detailed presentation of Fourier analysis in Chapter 3. We focus on ideas that have counterparts in wavelet theory, e.g., the approximation properties and the reflection of properties of a function in the expansion coefficients.

Chapters 4–5 deal with wavelets and their properties. Chapter 4 describes wavelets more in words rather than symbols, but it will give the reader an understanding of the fundamental questions and concepts involved. This chapter also tells the story of how the wavelet era began and discusses applications in signal processing. Chapter 5 is slightly more technical, but still at a level that is understandable based on the knowledge of the book's first chapters. A central part of this chapter is Section 5.4, which explains the multiscale representation associated to wavelets in the special case of the Haar wavelet.

Finally, Appendix A contains proofs of selected results, and Appendix B lists some important power series and Fourier series.

The list of references contains articles and books at several levels. In order to be more informative, we have introduced the following ranking system: **(A)** elementary; **(B)** undergraduate level; **(C)** graduate level; **(D)** research paper; **(H)** historical paper. The corresponding ranking of the book's contents is as follows:

(A): Chapter 1;

(A-B): Chapter 2, Chapter 4, appendix;

(B): Chapter 3, Chapter 5.

The presentation of the material on infinite series was inspired by notes used at the Technical University of Denmark. In their original version, these notes were written by H. E. Jensen; several professors from the Department of Mathematics have contributed to the later versions. Some of our examples are borrowed from these notes.

We would like to thank a number of mathematicians who have helped us in the process of writing this book: Preben Alsholm and Ole Jørsboe for pointing out mistakes in early drafts of the manuscript; Per Christian Hansen and Torben K. Jensen for their help with the figures; Palle Jorgensen and Jakob Lemvig for useful suggestions; and Albert Cohen and Mladen Victor Wickerhauser for providing special figures. We also thank John Benedetto, ANHA series editor, for proposing that we write this book, and Hans Feichtinger for providing us with office facilities during a long-term stay at NuHAG in Vienna. Finally, we thank the staff of Birkhäuser, especially Tom Grasso, for very efficient editorial assistance; and Elizabeth Loew from TeXniques, who helped with Latex problems.

The first-named author acknowledges support from the WAVE-program, sponsored by the Danish Science Foundation.

Ole Christensen
Khadija Laghrida Christensen
Kgs. Lyngby, Denmark
January 2004

In the present reprint, a small number of misprints are corrected. Example 3.2.6 is modified, and Exercises 2.13, 2.14, 2.15 and 3.10 are new.

Ole Christensen
Khadija Laghrida Christensen
Kgs. Lyngby, Denmark
December 2004

1

Approximation with Polynomials

In many applications of mathematics, we face functions which are far more complicated than the standard functions from classical analysis. Some of these functions can not be expressed in closed form via the standard functions, and some are only known implicitly or via their graph. Think, for example, of an electric circuit, where we are measuring the current at a certain point as a function of time: the outcome might be quite complicated, and best described via a graph.

An engineer measuring the current in an electric circuit will speak about having a *signal;* for a mathematician, this just means that the output of the measurement is a function, to be called f, for which $f(x)$ equals the current at the time x. We will not give an exact definition of what is meant by a signal: for our purpose, it is sufficient to think about a signal as a manifestation of a physical event in terms of a function, or, in cases to be discussed in Chapter 5, a sequence of numbers.

Signals are usually not given directly in terms of a function; for example, they often appear via a measurement. This makes it difficult or impossible to extract exact information on the signal from the function f describing it, especially if we need to perform some calculations on f. In such cases, it is important to be able to *approximate* f with a simpler function; that is, we would like to find a function, to be called g, for which

- the relevant calculation can be performed on the function g;
- the function g is close to f, in the sense that the outcome of the calculation performed on g gives useful information about the signal described by f.

In this chapter we focus on approximation with polynomials. In its simplest version, this idea appears already in the context of finding the tangent line for a differentiable function f at a point x_0; in fact, the tangent line at x_0 is a simple function, namely, a polynomial of degree 1, which usually approximates the function f quite well in a neighborhood of x_0.

We will show how better approximations can be obtained using polynomials of higher degree. We begin with a general introduction to the idea of approximation in Section 1.1. Then we move on with approximation by polynomials in Section 1.2, which is devoted to Weierstrass' theorem: it says that *any* continuous function on a closed and bounded interval can be approximated arbitrarily well by a polynomial. In Section 1.3 we obtain a more concrete statement in Taylor's theorem, which, under certain conditions, tells us how we can choose a polynomial which approximates a given function.

1.1 Approximation of a function on an interval

Our starting point must be a precise definition of what it means that a function is approximated by another function. As we will see soon, there are in fact several different ways of defining this, and the correct definition depends on the situation at hand. Let us give a concrete example where approximation theory is needed:

Example 1.1.1 Assume that we want to calculate the integral

$$\int_0^1 e^{-x^2}\,dx. \tag{1.1}$$

It is well known that there does not exist a formula for the integral $\int e^{-x^2}\,dx$ in terms of the elementary functions; that is, we can not find (1.1) simply by integrating the function e^{-x^2} and then inserting the limits $x = 0, x = 1$. Thus, we have to find another way to estimate (1.1). This is the point where approximation theory enters the picture: in this concrete case, our program on page 1 suggests that we search for a function g for which

- $\int_0^1 g(x)dx$ can be calculated, and
- $g(x)$ is close to e^{-x^2} for $x \in [0,1]$, in the sense that we can control how much $\int_0^1 g(x)dx$ deviates from $\int_0^1 e^{-x^2}\,dx$.

One way of doing so is to find a positive function g which we can integrate, and for which, for some $\epsilon > 0$,

$$-\epsilon \le e^{-x^2} - g(x) \le \epsilon, \ \forall x \in [0,1],$$

or

$$-\epsilon + g(x) \leq e^{-x^2} \leq \epsilon + g(x), \ \forall x \in [0,1].$$

This, in turn, implies that

$$\int_0^1 (-\epsilon + g(x))dx \leq \int_0^1 e^{-x^2}dx \leq \int_0^1 (\epsilon + g(x))dx,$$

and therefore

$$-\epsilon + \int_0^1 g(x)dx \leq \int_0^1 e^{-x^2}dx \leq \epsilon + \int_0^1 g(x)dx.$$

Thus, $\int_0^1 g(x)dx$ gives us an approximate value for the desired integral. After developing the necessary theory in Section 1.3, we come back to the question about how to chose g in Example 1.3.8. □

The argument in Example 1.1.1 can be generalized. In fact, exactly the same consideration leads to the following:

Proposition 1.1.2 *Let $I \subset \mathbb{R}$ be a finite interval with length L. Furthermore, let f and g be integrable functions defined on I, and assume that for some $\epsilon > 0$,*

$$|f(x) - g(x)| \leq \epsilon, \ \forall x \in I. \tag{1.2}$$

Then

$$-\epsilon L + \int_I g(x)dx \leq \int_I f(x)dx \leq \epsilon L + \int_I g(x)dx.$$

Thus, if our purpose is to approximate an integral of a complicated function f, the inequality (1.2) describes a property of the function g which makes it possible to estimate the integral of f (if we can find the integral of g).

In this chapter, we consequently consider approximation in the sense of (1.2); that is, we will consider a function g to be a good approximation of f if (1.2) is satisfied with a sufficiently small value of $\epsilon > 0$.

Usually, the word *uniform approximation* is associated with the particular condition in (1.2). It is a suitable definition of approximation for many practical purposes; however, there are also cases where the fact that g approximates f in the uniform sense does not allow us to extract the desired information. For example, the condition (1.2) does not lead to any information about the difference between the derivatives f' and g':

Example 1.1.3 Let $f(x) = \cos x$ and consider for $k \in \mathbb{N}$ the functions

$$g_k(x) = \cos x + \frac{1}{k}\sin(k^2 x).$$

Then

$$|f(x) - g_k(x)| \leq \frac{1}{k}, \ \forall x \in \mathbb{R};$$

that is, by choosing k sufficiently large, we can make g_k approximate f as well as we want in the uniform sense. On the other hand, since

$$f'(x) = -\sin x, \quad g_k'(x) = -\sin x + k\cos(k^2 x),$$

we see that

$$|f'(x) - g_k'(x)| = k\,|\cos(k^2 x)|.$$

This implies that

$$\sup_{x\in\mathbb{R}} |f'(x) - g_k'(x)| = |f'(0) - g_k'(0)| = k.$$

Thus, despite the fact that g_k converges to f in the uniform sense, the distance between f' and g_k' is increasing. □

This example shows that if our purpose is to approximate the derivative of a function, a different definition of approximation is needed; uniform approximation is not suitable.

1.2 Weierstrass' theorem

We will now consider approximation in the sense of (1.2), with the extra requirement that we want g to be a polynomial. That is, given a function f defined on an interval I, we ask whether we, for a given $\epsilon > 0$, can find a polynomial P such that

$$|f(x) - P(x)| \leq \epsilon \text{ for all } x \in I.$$

A general polynomial can be written in the form

$$P(x) = a_N x^N + a_{N-1}x^{N-1} + \cdots + a_2 x^2 + a_1 x + a_0 = \sum_{n=0}^{N} a_n x^n,$$

where $a_0, \ldots, a_N$ are some numbers; if $a_N \neq 0$, the polynomial has *degree* N. It turns out that the answer to our question might be no:

Example 1.2.1 Consider the function $f :]-1,1[\to \mathbb{R}$ given by

$$f(x) = \begin{cases} 1 & \text{for } x \in]-1,0[, \\ 0 & \text{for } x \in [0,1[, \end{cases}$$

and let $\epsilon = 0.1$. Then there is no polynomial P for which

$$|f(x) - P(x)| \leq 0.1 \text{ for all } x \in]-1,1[. \tag{1.3}$$

In fact, if there was such a polynomial P, then for all $x \in]-1,0[$,

$$P(x) \geq f(x) - 0.1 \geq 0.9.$$

Since every polynomial is continuous, it follows that

$$P(0) = \lim_{x \to 0} P(x) \geq 0.9.$$

But $f(0) = 0$, so this implies that

$$|f(0) - P(0)| = |P(0)| \geq 0.9.$$

This contradicts (1.3); we conclude that no polynomial can satisfy (1.3). □

Precisely the same argument shows that we can never approximate a discontinuous function arbitrarily well by a polynomial: a single point of discontinuity is enough to make this impossible. For this reason we will only consider continuous functions in this chapter. A surprising result is that every continuous function on a closed and bounded interval can be approximated arbitrarily well with a polynomial. This is the famous *Weierstrass' theorem:*

Theorem 1.2.2 *Let $I \subset \mathbb{R}$ be a closed and bounded interval and f a continuous function defined on I. Then, for every $\epsilon > 0$ there exists a polynomial P such that*

$$|f(x) - P(x)| \leq \epsilon \text{ for all } x \in I. \tag{1.4}$$

The theorem is proved in Section A.2. The geometric meaning of the theorem is that if we surround the graph of f with a band having width 2ϵ for some $\epsilon > 0$ (see Figure 1.3.1), then there exists a polynomial that goes completely inside this band. The interesting thing about the theorem is that this is possible *regardless how little ϵ is chosen, i.e., regardless how narrow the band is.* Imagine for example how it will look like if $\epsilon = 10^{-10}$! Continuous functions can easily "behave wildly" (signals from a seismograph, medical instruments), so this is really a strong statement. See, e.g., the speech signal in Figure 1.3.2; intuitively, it is hard to imagine a polynomial which is close to such a signal over the entire interval, but Weierstrass' theorem says that such a polynomial exists.

The approximating polynomial P in (1.4) will in general depend on at least three factors, namely,

- how well we want f approximated, i.e., how small ϵ is;
- the behavior of f; strong oscillations in f usually force P to be of a high degree;
- the length of the interval I; enlarging the interval in general forces us to choose polynomials of higher degree if a certain approximation has to be obtained.

For example, it can be that for $\epsilon = 1$ we can approximate f with a second degree polynomial

$$P(x) = a_2x^2 + a_1x + a_0,$$

while for $\epsilon = 0.1$ we might need to choose a polynomial of degree 10,

$$P(x) = a_{10}x^{10} + a_9x^9 + \cdots + a_1x + a_0.$$

In principle Weierstrass' theorem makes it possible to replace complicated functions by approximating polynomials in many situations. In practice, however, it is a problem that this is only an *existence theorem:* it does not say how one can choose the polynomial that approximates a given function. The proof partly compensates for this, but the theorem is still hard to use in practice. Under extra assumptions we can reach a much more useful result, namely *Taylor's theorem.*

1.3 Taylor's theorem

In order to speak about the tangent line of a function f at a point x_0 we need to assume that the function f is differentiable at x_0. The tangent line usually approximates f well in a small neighborhood of x_0; for this reason, the tangent line is frequently called the ***approximating polynomial of degree*** 1. Intuitively, it is very reasonable that we should be able to obtain better approximations using polynomials of higher order. If we want to approximate f via a polynomial of second degree, it turns out to be very convenient to assume also that f' is differentiable at x_0, i.e., that f is twice differentiable at x_0. In Taylor's theorem we will approximate with polynomials of degree N for some $N \in \mathbb{N}$, and in this context we need to assume that all the derivatives up to order N exist. It turns out that there is a close link between how well we want our function f to be approximated, and the order of the needed polynomials. If we want to obtain a sequence of polynomials approximating f better and better, this means that we need to assume that f is ***arbitrarily often differentiable*** at the point x_0. The nth derivative of the function f at x_0 will be denoted by $f^{(n)}(x_0)$; for $n = 1, 2$ we will also use the notation $f'(x_0), f''(x_0)$.

In Taylor's theorem we will approximate the function f via polynomials P_N, $N \in \mathbb{N}$, of the form

$$\begin{aligned} &P_N(x) \\ &= f(x_0) + \frac{f'(x_0)}{1!}(x - x_0) + \frac{f''(x_0)}{2!}(x - x_0)^2 + \cdots + \frac{f^{(N)}(x_0)}{N!}(x - x_0)^N \\ &= \sum_{n=0}^{N} \frac{f^{(n)}(x_0)}{n!}(x - x_0)^n. \end{aligned}$$

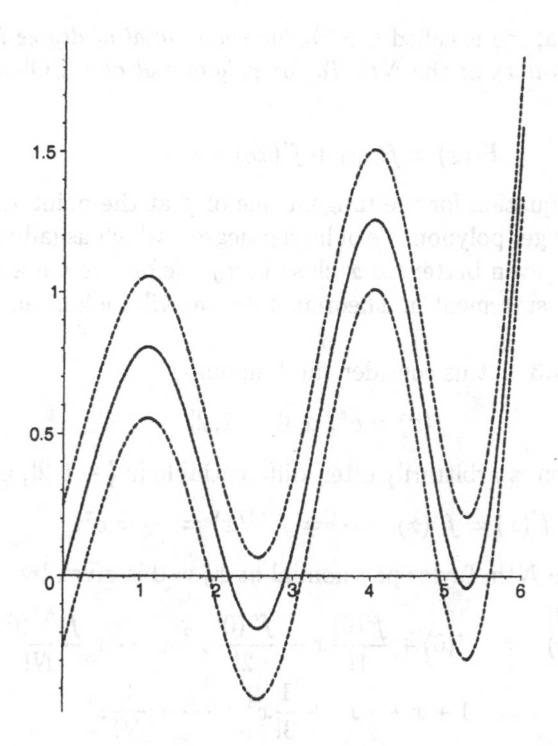

Figure 1.3.1 *A function f, (unbroken line) together with the functions $f+\epsilon, f-\epsilon$, which give an ϵ-band around the graph of f. According to Theorem 1.2.2 there exists a polynomial that goes inside this band, no matter how small ϵ is chosen, i.e., no matter how narrow the band is.*

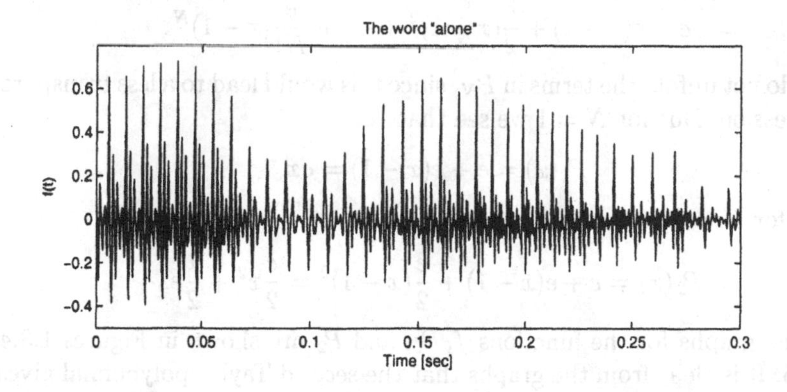

Figure 1.3.2 *A speech signal. One can think about such a signal as the current in the cable to the loudspeaker when a recording of the speech is played. The actual signal is a recording of the word "alone".*

The polynomial P_N is called the *Taylor polynomial of degree N associated to f at the point x_0* or the *Nth Taylor polynomial at x_0*. Observe that for $N = 1$,

$$P_1(x) = f(x_0) + f'(x_0)(x - x_0),$$

which is the equation for the tangent line of f at the point x_0. For higher values of N we get polynomials of higher degree, which usually approximate the function f even better for x close to x_0. Before we see a more precise version of this statement in Theorem 1.3.6 we will look at an example.

Example 1.3.3 Let us consider the function

$$f(x) = e^x, \ x \in]-1, 2[.$$

This function is arbitrarily often differentiable in $]-1, 2[$, and

$$f'(x) = f''(x) = \cdots = f^{(n)}(x) = \cdots = e^x.$$

For $N \in \mathbb{N}$ the Nth Taylor polynomial at $x_0 = 0$ is given by

$$\begin{aligned} P_N(x) &= f(0) + \frac{f'(0)}{1!}x + \frac{f''(0)}{2!}x^2 + \cdots + \frac{f^{(N)}(0)}{N!}x^N \\ &= 1 + x + \frac{1}{2}x^2 + \frac{1}{3!}x^3 + \cdots + \frac{1}{N!}x^N \\ &= \sum_{n=0}^{N} \frac{1}{n!}x^n. \end{aligned} \tag{1.5}$$

Similarly, the Nth Taylor polynomial at $x_0 = 1$ is given by

$$\begin{aligned} P_N(x) &= f(1) + \frac{f'(1)}{1!}(x-1) + \frac{f''(1)}{2!}(x-1)^2 + \cdots + \frac{f^{(N)}(1)}{N!}(x-1)^N \\ &= e + e(x-1) + \frac{e}{2}(x-1)^2 + \cdots + \frac{e}{N!}(x-1)^N. \end{aligned}$$

We do not unfold the terms in P_N, since this would lead to a less transparent expression. But for $N = 1$ we see that

$$P_1(x) = e + e(x-1) = ex,$$

and for $N = 2$,

$$P_2(x) = e + e(x-1) + \frac{e}{2}(x-1)^2 = \frac{e}{2}x^2 + \frac{1}{2}e.$$

The graphs for the functions f, P_1, and P_2 are shown in Figures 1.3.4 – 1.3.5. It is clear from the graphs that the second Taylor polynomial gives a better approximation of f around $x = 1$ than the first Taylor polynomial. One can reach an even better approximation by choosing a Taylor polynomial of higher degree, which however requires more calculations. In the present example it is easy to find the Taylor polynomials of higher degree; in real applications however, there will always be a tradeoff between how

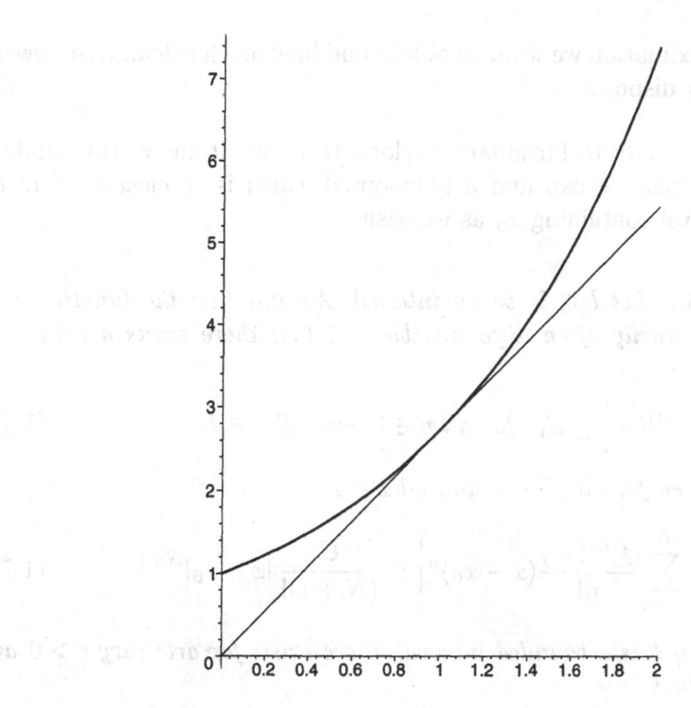

Figure 1.3.4 *The function $f(x) = e^x$ together with the first Taylor polynomial at $x_0 = 1$ (thin line).*

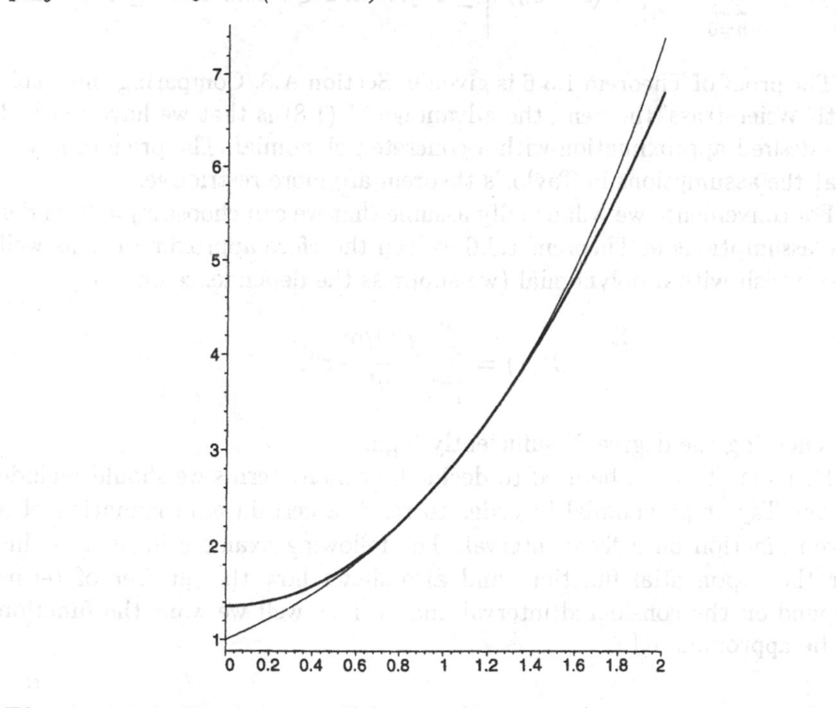

Figure 1.3.5 *The function $f(x) = e^x$ (thin curve) together with the second Taylor polynomial at $x_0 = 1$.*

good an approximation we want to obtain and how much calculation power we have at our disposal. □

We are now ready to formulate Taylor's theorem. It shows that under certain conditions, we can find a polynomial which is as close to f in a bounded interval containing x_0 as we wish:

Theorem 1.3.6 *Let $I \subset \mathbb{R}$ be an interval. Assume that the function $f : I \to \mathbb{R}$ is arbitrarily often differentiable and that there exists a constant $C > 0$ such that*

$$\left|f^{(n)}(x)\right| \leq C, \quad \text{for all } n \in \mathbb{N} \text{ and all } x \in I. \tag{1.6}$$

Let $x_0 \in I$. Then for all $N \in \mathbb{N}$ and all $x \in I$,

$$\left| f(x) - \sum_{n=0}^{N} \frac{f^{(n)}(x_0)}{n!}(x - x_0)^n \right| \leq \frac{C}{(N+1)!} |x - x_0|^{N+1}. \tag{1.7}$$

In particular, if I is a bounded interval, there exists for arbitrary $\epsilon > 0$ an $N_0 \in \mathbb{N}$ such that

$$\left| f(x) - \sum_{n=0}^{N} \frac{f^{(n)}(x_0)}{n!}(x - x_0)^n \right| \leq \epsilon \quad \text{for all } x \in I \text{ and all } N \geq N_0. \tag{1.8}$$

The proof of Theorem 1.3.6 is given in Section A.3. Comparing the result with Weierstrass' theorem, the advantage of (1.8) is that we have reached the desired approximation with a concrete polynomial. The price to pay is that the assumptions in Taylor's theorem are more restrictive.

For convenience we will usually assume that we can choose $x_0 = 0$; under the assumptions in Theorem 1.3.6 we can therefore approximate f as well as we wish with a polynomial (we suppress the dependence on N)

$$P(x) = \sum_{n=0}^{N} \frac{f^{(n)}(0)}{n!} x^n,$$

by choosing the degree N sufficiently high.

Formula (1.7) can be used to decide how many terms we should include in the Taylor polynomial in order to reach a certain approximation of a given function on a fixed interval. The following example illustrates this for the exponential function, and also shows how the number of terms depend on the considered interval and on how well we want the function to be approximated.

Example 1.3.7 We consider again the function $f(x) = e^x$. We aim at finding three values of $N \in \mathbb{N}$, namely

(i) $N \in \mathbb{N}$ such that

$$\left| e^x - \sum_{n=0}^{N} \frac{x^n}{n!} \right| \leq 0.015 \quad \text{for all } x \in [-1,2]. \tag{1.9}$$

(ii) $N \in \mathbb{N}$ such that

$$\left| e^x - \sum_{n=0}^{N} \frac{x^n}{n!} \right| \leq 0.0015 \quad \text{for all } x \in [-1,2]. \tag{1.10}$$

(iii) $N \in \mathbb{N}$ such that

$$\left| e^x - \sum_{n=0}^{N} \frac{x^n}{n!} \right| \leq 0.015 \quad \text{for all } x \in [-5,5]. \tag{1.11}$$

In order to do so, we first observe that the estimate (1.7) involves the constant C, which we can choose as the maximum of all derivatives of f on the considered interval. For the function f we have that $f^{(n)}(x) = e^x$, i.e., all derivatives equal the function itself.

For (i) we can thus take $C = e^2$. Now (1.7) shows that (1.9) is obtained if we choose N such that

$$\frac{e^2}{(N+1)!} 2^{N+1} \leq 0.015; \tag{1.12}$$

this is satisfied for all $N \geq 8$.

In (ii), we obtain an appropriate value for $N \in \mathbb{N}$ by replacing the number 0.015 on the right-hand side of (1.12) by 0.0015; the obtained inequality is satisfied for all $N \geq 10$.

Note than in (iii), we enlarge the interval on which we want to approximate f. In order to find an appropriate N-value in (iii), we need to replace the previous value for C with $C = e^5$; we obtain the inequality

$$\frac{e^5}{(N+1)!} 5^{N+1} \leq 0.015,$$

which is satisfied for $N \geq 19$. □

We can use the result in Example 1.3.7 to estimate the integral in Example 1.1.1:

Example 1.3.8 We return to the question of estimation of

$$\int_0^1 e^{-x^2} dx.$$

When $x \in [0,1]$, we have that $-x^2 \in [-1,0]$; thus, if we replace x by $-x^2$ in (1.9), the result in Example 1.3.7(i) tells us that

$$\left| e^{-x^2} - \sum_{n=0}^{8} \frac{(-x^2)^n}{n!} \right| \leq 0.015 \quad \text{for all } x \in [0,1].$$

By Proposition 1.1.2 this implies that

$$-0.015+\int_0^1 \sum_{n=0}^{8} \frac{(-x^2)^n}{n!}dx \le \int_0^1 e^{-x^2}dx \le 0.015+\int_0^1 \sum_{n=0}^{8} \frac{(-x^2)^n}{n!}dx.$$

Now,

$$\begin{aligned}\int_0^1 \sum_{n=0}^{8} \frac{(-x^2)^n}{n!}dx &= \left[\sum_{n=0}^{8}(-1)^n \frac{x^{2n+1}}{(2n+1)n!}\right]_0^1 \\ &= \sum_{n=0}^{8}(-1)^n \frac{1}{(2n+1)n!} \\ &= 0.747. \end{aligned} \tag{1.13}$$

Taking the small round-off error in (1.13) into consideration, this shows that

$$\int_0^1 e^{-x^2}dx = 0.747 \pm 0.016. \tag{1.14}$$

The obtained result is actually much closer to the exact value of the integral than the above tolerance indicates. For $N = 8$, the inequality (1.9) holds for all $x \in [-1, 2]$; in the arguments leading to (1.14) we only used that the inequality (1.9) holds for all $x \in [-1, 0]$, and the reader can check that this is the case already for $N = 5$ (Exercise 1.3). Using $N = 8$ rather than $N = 5$ leads to a better approximation of the exponential function, and therefore to an approximation of the integral which is better than claimed in (1.14). See Exercise 1.3 and Example 2.4.3. □

When dealing with mathematical theorems, it is essential to understand the role of the various conditions; in fact, "mathematics" can almost be defined as the science of finding the minimal condition leading to a certain conclusion. For a statement like Theorem 1.3.6, this means that we should examine the conditions once more and find out whether they are really needed. Formulated more concretely in this particular case: is it really necessary that f is infinitely often differentiable on I and that (1.6) holds, in order to reach the conclusion in (1.7)?

This discussion would take us too far at the moment, but we can indicate the answer. In order for the conclusion in (1.7) to make sense for all $n \in \mathbb{N}$, we at least need f to be arbitrarily often differentiable at x_0. Furthermore, in Section 2.4 we will see that (1.7) leads to a so-called power series representation of f in a neighborhood of x_0 (i.e., for x sufficiently close to x_0); and Theorem 2.4.10 will show us that this forces f to be arbitrarily often differentiable in that neighborhood. What remains is a discussion concerning the necessity of (1.6). For this, we refer to our later Example 2.4.12. This shows that there exists a nonconstant function $f : \mathbb{R} \to \mathbb{R}$ for which

$$f \text{ is arbitrarily often differentiable and } f^{(n)}(0) = 0 \text{ for all } n \in \mathbb{N}.$$

Taking $x_0 = 0$, for such a function we have

$$f(x) - \sum_{n=0}^{N} \frac{f^{(n)}(x_0)}{n!}(x - x_0)^n = f(x) - f(0) \text{ for all } x \in \mathbb{R}.$$

This shows that (1.7) can not hold for all $N \in \mathbb{N}$ (otherwise we would have $f(x) = f(0)$ for all $x \in \mathbb{R}$). The reason that this does not contradict Theorem 1.3.6 is that the function in Example 2.4.12 does not satisfy (1.6). This shows that the assumption (1.6) actually plays a role in the theorem; it is not just added for convenience of the proof, the conclusion might not hold if it is removed.

Example 1.3.7 shows that the better approximation we need, the higher the degree of the approximating polynomial P has to be. There is nothing that is called a "polynomial of infinite degree", but one can ask if one can obtain that $f(x) = P(x)$ if one has the right to include "infinitely many terms" in $P(x)$. So far we have not defined a sum of infinitely many numbers, but our next purpose is to show that certain functions actually can be represented in the form

$$f(x) = a_0 + a_1x + a_2x^2 + \cdots + a_nx^n + \cdots = \sum_{n=0}^{\infty} a_nx^n. \tag{1.15}$$

At the moment it is far from being clear how such an infinite sum should be interpreted. In the next chapter we give the formal definition, and consider functions f which can be represented this way.

1.4 Exercises

1.1 Find the first three Taylor polynomials at $x_0 = 0$ for the function

$$f(x) = e^x \cos x + \ln(x + \frac{1}{2}),$$

and plot their graphs together with the graph of f.

1.2 Find the Nth Taylor polynomial at $x_0 = 0$ for the function

$$f(x) = \ln(1 + x).$$

1.3 (i) Find $N \in \mathbb{N}$ such that

$$\left| e^x - \sum_{n=0}^{N} \frac{x^n}{n!} \right| \leq 0.015 \text{ for all } x \in [-1, 0].$$

(ii) Show that

$$\left| e^x - \sum_{n=0}^{8} \frac{x^n}{n!} \right| \leq 2.8 \cdot 10^{-6}, \quad \text{for all } x \in [-1, 0].$$

(iii) Show that

$$\int_0^1 e^{-x^2} dx = 0.746824 \pm 4 \cdot 10^{-6}.$$

1.4 (i) Find a polynomial $P(x) = \sum_{n=0}^{N} a_n x^n$ such that

$$\left| \sin x - \sum_{n=0}^{N} a_n x^n \right| \leq 0.1, \quad \forall x \in [0, \frac{\pi}{2}].$$

(ii) Use the result to find an approximative value of

$$\int_0^1 \sin x^4 \, dx,$$

with an error of at most 0.1.

2

Infinite Series

Motivated by (1.15) at the end of the previous chapter, we wish to define an infinite sum of power functions; however, before we can do so, we need to define an infinite sum of real (or complex) numbers. That is, our first task will be to consider an infinite sequence of numbers $a_1, a_2, \ldots, a_n, \ldots$, and examine when and how we can make sense of the sum

$$a_1 + a_2 + a_3 + \cdots + a_n + \cdots .$$

The formal definition of adding infinitely many numbers will be given in Section 2.1. Compared with addition of finitely many numbers, it turns out that many complications appear. For example, the sum might depend on the way the numbers are ordered; and in many cases, only approximative values for the sum can be found. Section 2.1 also introduces some of the criteria which are used to decide whether a given infinite summation makes sense. Section 2.2 describes some methods for estimating infinite sums. After these preliminary steps, we turn to a subject which will follow us during the entire book, namely, infinite sums with terms depending on a variable. We analyze such cases with increasing complexity: in Section 2.3 we consider geometric series, i.e., sums of the type (x denotes the variable)

$$1 + x + x^2 + \cdots + x^n + \cdots . \tag{2.1}$$

In Section 2.4 we consider "similar" series, where we allow the terms in (2.1) to be multiplied by some constants, i.e., the series are of the form

$$a_0 + a_1x + a_2x^2 + \cdots + a_nx^n + \cdots$$

for certain constants a_n; series of this type are called *power series.* Section 2.5 goes one important step further: we consider infinite sums of general functions, defined on some interval. An understanding of this subject is instrumental for the analysis of Fourier series and wavelets. Section 2.6 introduces uniform convergence, and finally, Section 2.7 gives an idea about how infinite series can be useful for signal transmission.

In the entire chapter we focus on the approximation properties related to infinite series. That is, given an infinite series of numbers, how can we estimate how many terms we need to include in order to obtain a realistic approximation of the sum? And given an infinite series representing a function, how can we guarantee that the finite partial sums share the properties of the function?

2.1 Infinite series of numbers

We begin by defining an infinite sum of real (or complex) numbers $a_1, a_2, a_3, \ldots, a_n, \ldots$.

Formally, an *infinite series* is an expression of the form

$$a_1 + a_2 + a_3 + \cdots + a_n + \cdots = \sum_{n=1}^{\infty} a_n.$$

The numbers a_n are called the *terms* in the infinite series. In order to explain how an infinite series should be interpreted, we first define the Nth *partial sum* associated to the infinite series as

$$S_N := a_1 + a_2 + a_3 + \cdots + a_N = \sum_{n=1}^{N} a_n. \tag{2.2}$$

Definition 2.1.1 *Let* $\sum_{n=1}^{\infty} a_n$ *be an infinite series with partial sums* S_N. *If* S_N *is convergent, i.e., if there exists a number* S *such that*

$$S_N \to S \text{ as } N \to \infty, \tag{2.3}$$

we say that the infinite series $\sum_{n=1}^{\infty} a_n$ *is convergent with sum* S. *We write this as*

$$\sum_{n=1}^{\infty} a_n = S.$$

In case S_N *is divergent,* $\sum_{n=1}^{\infty} a_n$ *is said to be divergent. No sum is attributed to a divergent series.*

Note that if we express the definition directly via the terms in the infinite series, to say that $\sum_{n=1}^{\infty} a_n$ is convergent with sum S means exactly that

$$\sum_{n=1}^{N} a_n \to S \text{ as } N \to \infty;$$

thus, to consider an infinite sum really corresponds to the intuitive idea of just "adding more and more terms".

To avoid confusion, we mention that sometimes we will consider infinite series starting with a_0 instead of a_1. The difference between $\sum_{n=1}^{\infty} a_n$ and $\sum_{n=0}^{\infty} a_n$ is that in the second case we have added the number a_0; so for the question whether the series is convergent or not, it does not matter whether we start with $n = 0$ or $n = 1$. In case of convergence,

$$\sum_{n=0}^{\infty} a_n = a_0 + \sum_{n=1}^{\infty} a_n.$$

To determine whether a series is convergent or not, one usually applies the so-called convergence criteria, to which we return later in this section. In some cases we can apply the definition directly:

Example 2.1.2 In this example we consider the infinite series

(i) $\sum_{n=1}^{\infty} n = 1 + 2 + \cdots + n + \cdots;$

(ii) $\sum_{n=1}^{\infty} \frac{1}{2^n} = \frac{1}{2} + \frac{1}{2^2} + \cdots + \frac{1}{2^n} + \cdots;$

(iii) $\sum_{n=1}^{\infty} \frac{1}{n(n+1)} = \frac{1}{1\cdot 2} + \frac{1}{2\cdot 3} + \cdots + \frac{1}{n(n+1)} + \cdots.$

For the infinite series in (i), the Nth partial sum is

$$S_N = \sum_{n=1}^{N} n = 1 + 2 + \cdots + N.$$

It is clear that

$$S_N \to \infty \text{ for } N \to \infty$$

(in fact, $S_N = \frac{N(N+1)}{2}$ for all $N \in \mathbb{N}$). Thus

$$\sum_{n=1}^{\infty} n \text{ is divergent.}$$

Now we consider the infinite series in (ii). The Nth partial sum is

$$S_N = \frac{1}{2} + \frac{1}{2^2} + \cdots + \frac{1}{2^N} = \sum_{n=1}^{N} \frac{1}{2^n}.$$

We now give a geometric argument showing that

$$\sum_{n=1}^{N} \frac{1}{2^n} = 1 - \frac{1}{2^N}. \tag{2.4}$$

In fact, $S_1 = \frac{1}{2}$, which is exactly half of the distance from 0 to 1. That is,

$$S_1 = 1 - \frac{1}{2}.$$

Further, $S_2 = \frac{1}{2} + \frac{1}{2^2}$; thus, the extra contribution compared to S_1 is $\frac{1}{2^2}$, which is half the distance from S_1 to 1. Going to the next step, i.e., looking at S_3, again cuts the distance to 1 by a factor of 2:

$$S_3 = \frac{1}{2} + \frac{1}{2^2} + \frac{1}{2^3} = 1 - \frac{1}{2^3}.$$

The same happens in all the following steps: when looking at S_N, i.e., after N steps, the distance to 1 will be exactly $1/2^N$, which shows (2.4). It follows that

$$\sum_{n=1}^{N} \frac{1}{2^n} \to 1 \text{ for } N \to \infty;$$

we conclude that

$$\sum_{n=1}^{\infty} \frac{1}{2^n} \text{ is convergent with sum 1.}$$

Concerning the series in (iii), we first observe that for any $n \in \mathbb{N}$,

$$\frac{1}{n(n+1)} = \frac{1}{n} - \frac{1}{n+1};$$

thus, the Nth partial sum is

$$\begin{aligned} S_N &= \frac{1}{1\cdot 2} + \frac{1}{2\cdot 3} + \cdots + \frac{1}{N(N+1)} \\ &= \left(1 - \frac{1}{2}\right) + \left(\frac{1}{2} - \frac{1}{3}\right) + \cdots + \left(\frac{1}{N} - \frac{1}{N+1}\right) \\ &= 1 - \frac{1}{N+1}. \end{aligned}$$

Since $S_N \to 1$ as $N \to \infty$, we conclude that the series is convergent with sum 1. □

Intuitively, the "problem" with the series in Example 2.1.2 (i) is that the terms $a_n = n$ are too large: considering partial sums S_N for larger and larger values of N means that we continue to add numbers to S_N, and when they grow as in this example, it prevents S_N from having a finite limit as $N \to \infty$. A refinement of this intuitive statement leads to a useful criterion, which states that an infinite series only has a chance to be convergent if $a_n \to 0$ as $n \to \infty$; this result is called the *nth term test.*

Theorem 2.1.3 *If $a_n \nrightarrow 0$ as $n \to \infty$, then $\sum_{n=1}^{\infty} a_n$ is divergent.*

Proof: Assume that $\sum_{n=1}^{\infty} a_n$ is convergent with sum S. By definition of the partial sums, this means that

$$S_N = a_0 + a_1 + \cdots + a_N \to S \text{ as } N \to \infty;$$

this implies that

$$a_N = S_N - S_{N-1} \to S - S = 0 \text{ as } N \to \infty. \qquad \square$$

An alternative (but equivalent) formulation of Theorem 2.1.3 says that

$$\text{if } \sum_{n=1}^{\infty} a_n \text{ is convergent, then } a_n \to 0 \text{ for } n \to \infty.$$

It is important to notice that Theorem 2.1.3 (at most) can give us a negative conclusion: we can *never* use the nth term test to conclude that a series is convergent. In fact, even if $a_n \to 0$ as $n \to \infty$, it can happen that $\sum_{n=1}^{\infty} a_n$ is divergent; see Example 2.2.3.

Intuitively, it is very reasonable that if a certain infinite series with positive terms is convergent, then any series with smaller (positive) terms is also convergent; in fact, the partial sums of the series with the smaller terms are forced to stay bounded. The formal statement of this result is called the *comparison test:*

Theorem 2.1.4 *Assume that $0 \le a_n \le b_n$ for all $n \in \mathbb{N}$. Then the following holds:*

(i) If $\sum_{n=1}^{\infty} b_n$ is convergent, then $\sum_{n=1}^{\infty} a_n$ is convergent as well.

(ii) If $\sum_{n=1}^{\infty} a_n$ is divergent, then $\sum_{n=1}^{\infty} b_n$ is divergent as well.

Example 2.1.5 Consider the infinite series

$$\sum_{n=1}^{\infty} \frac{1}{(n+1)^2} = \frac{1}{2^2} + \frac{1}{3^2} + \cdots + \frac{1}{(n+1)^2} + \cdots .$$

First, observe that for all $n \in \mathbb{N}$,

$$\frac{1}{(n+1)^2} \le \frac{1}{(n+1)n}.$$

By Example 2.1.2(iii) we know that $\sum_{n=1}^{\infty} \frac{1}{(n+1)n}$ is convergent, so the comparison criterion shows that $\sum_{n=1}^{\infty} \frac{1}{(n+1)^2}$ is convergent. This coincides with a result we will obtain much later in Example 3.6.2, with a very different argument which even allows us to find the sum. $\square$

For the question whether a given series $\sum_{n=1}^{\infty} a_n$ is convergent or not, the first terms in the series do not matter: the question is whether a_n converges sufficiently fast to zero as $n \to \infty$. It turns out that if two series "behave similarly" for large values of n, they are either both convergent, or both divergent. In order to state this result formally, we need a definition: we say that two series $\sum_{n=1}^{\infty} a_n$ and $\sum_{n=1}^{\infty} b_n$ with positive terms are *equivalent* if there exists a constant $C > 0$ such that

$$\frac{a_n}{b_n} \to C \text{ for } n \to \infty.$$

Proposition 2.1.6 *Assume that the series* $\sum_{n=1}^{\infty} a_n$ *and* $\sum_{n=1}^{\infty} b_n$ *have positive terms and are equivalent. Then* $\sum_{n=1}^{\infty} a_n$ *and* $\sum_{n=1}^{\infty} b_n$ *are either both convergent, or both divergent.*

Proposition 2.1.6 gives an alternative argument for the result in Example 2.1.5: the reader can check that the series $\sum_{n=1}^{\infty} \frac{1}{(n+1)n}$ and $\sum_{n=1}^{\infty} \frac{1}{(n+1)^2}$ are equivalent.

Given a sequence of numbers, we note that our definition of the associated infinite series depends on how the terms are ordered: different orderings will lead to different partial sums, and (potentially) different answers to the question whether the series is convergent or not. This is not merely a theoretical concern, but a problem that appears in practice: in fact, there exist sequences where certain orderings lead to convergent series, and other lead to divergent series (see below). The following definition introduces a condition on the terms in an infinite series, which implies that the series is convergent with respect to any ordering.

Definition 2.1.7 *An infinite series* $\sum_{n=1}^{\infty} a_n$ *is said to be absolutely convergent if* $\sum_{n=1}^{\infty} |a_n|$ *is convergent.*

Proposition 2.1.8 *If* $\sum_{n=1}^{\infty} a_n$ *is absolutely convergent, then* $\sum_{n=1}^{\infty} a_n$ *is convergent; furthermore, any reordering of the terms in the series will lead to a convergent series having the same sum.*

A series which is convergent no matter how the terms are reordered, is said to be *unconditionally convergent.* Using this terminology, Proposition 2.1.8 says that an absolutely convergent series is unconditionally convergent.

There exist series which are convergent, but not absolutely convergent; they are said to be *conditionally convergent.* For conditionally convergent series with real terms, something very strange happens. In fact, the terms can be reordered such that we obtain a divergent series; and for *any* given real number $\tilde{S}$ we can find an ordering so that the corresponding infinite series is convergent with sum equal to $\tilde{S}$. Example 2.2.5 exhibits an infinite series which is convergent, but not absolutely convergent.

A sufficient condition for (absolute) convergence of an infinite series is given by the *quotient test*:

Theorem 2.1.9 *Assume that $a_n \neq 0$ for all $n \in \mathbb{N}$ and that*

$$\left|\frac{a_{n+1}}{a_n}\right| \to C \in [0,\infty] \quad \textit{for} \quad n \to \infty. \tag{2.5}$$

Then the following hold:

(i) If $C < 1$, then $\sum a_n$ is absolutely convergent.

(ii) If $C > 1$, then $\sum a_n$ is divergent.

We prove Theorem 2.1.9 in Section A.4. Note that the case where (2.5) is satisfied with $C = 1$ has a very special status: Theorem 2.1.9 does not give any conclusion in that case. Example 2.2.3 will show us that there exist convergent as well as divergent series which lead to $C = 1$; so if we end up with this particular value, we have to use other methods to find out whether the given series is convergent or not.

Example 2.1.10 Consider the infinite series

$$\sum_{n=1}^{\infty} \frac{n^2}{2^n} = \frac{1}{2} + 1 + \frac{9}{8} + \cdots + \frac{n^2}{2^n} + \cdots.$$

Letting $a_n = \frac{n^2}{2^n}$, we see that

$$\left|\frac{a_{n+1}}{a_n}\right| = \frac{(n+1)^2}{2^{n+1}} \cdot \frac{2^n}{n^2} = \left(1+\frac{1}{n}\right)^2 \frac{1}{2} \to \frac{1}{2} \quad \text{for } n \to \infty.$$

Thus the quotient test shows that the series is convergent. □

2.2 Estimating the sum of an infinite series

The definition of an infinite series involves the sum of the series. There exist several ways to check whether a series is convergent or not, but only in relatively rare cases is it possible to find the exact sum of an infinite series. This raises the question how one can find approximative values for the infinite sum. Using computers, we are able to calculate the partial sums S_N for large values of $N \in \mathbb{N}$; but in order for this information to be useful, we need to be able to estimate the difference

$$\left|\sum_{n=1}^{\infty} a_n - S_N\right| = \left|\sum_{n=1}^{\infty} a_n - \sum_{n=1}^{N} a_n\right| = \left|\sum_{n=N+1}^{\infty} a_n\right|.$$

One important result which can frequently be used in order to do so can be derived from the following result, called the *integral test:*

Theorem 2.2.1 *Assume that* $f : [1,\infty[\to [0,\infty[$ *is continuous and decreasing. Then the following holds:*

(i) If $\int_1^t f(x)dx$ *has a finite limit as* $t\to\infty$, *then* $\sum_{n=1}^\infty f(n)$ *is convergent, and*

$$\int_1^\infty f(x)dx < \sum_{n=1}^\infty f(n) < \int_1^\infty f(x)dx + f(1).$$

(ii) If $\int_1^t f(x)dx \to \infty$ *as* $t\to\infty$, *then* $\sum_{n=1}^\infty f(n)$ *is divergent.*

The theorem is proved in Section A.4. A reader who checks the proof will observe that it does not matter that the sum appearing in the integral test starts with $n = 1$; the same argument works starting with any positive integer-value of n. One way of formulating this observation is that if the conditions in Theorem 2.2.1 are satisfied, then for any $p \in \mathbb{N}$,

$$\int_p^\infty f(x)dx < \sum_{n=p}^\infty f(n) < \int_p^\infty f(x)dx + f(p). \tag{2.6}$$

We also notice that this result is obtained, even if f only satisfies the conditions in Theorem 2.2.1 starting with $x = p$. This leads to the following way of estimating the infinite sum $\sum_{n=1}^\infty f(n)$:

Corollary 2.2.2 *Assume that the function* $f : [1,\infty[\to [0,\infty[$ *is continuous and decreasing, at least from* $x = p$ *for some* $p \in \mathbb{N}$. *Then, assuming that* $\int_p^t f(x)dx$ *has a finite limit as* $t\to\infty$,

$$\left|\sum_{n=1}^\infty f(n) - \sum_{n=1}^N f(n)\right| \le \int_{N+1}^\infty f(x)dx + f(N+1)$$

for all integers $N \ge p$.

Corollary 2.2.2 is proved using that

$$\sum_{n=1}^\infty f(n) - \sum_{n=1}^N f(n) = \sum_{n=N+1}^\infty f(n),$$

followed by an application of (2.6) with $p = N+1$. In the following example, this result is used to estimate an infinite sum:

Example 2.2.3 Let $\alpha \ge 1$, and consider the function

$$f(x) = \frac{1}{x^\alpha},\ x \in [1,\infty[.$$

This function is decreasing and continuous. If $\alpha > 1$, then

$$\int_1^t \frac{1}{x^\alpha}dx = \frac{1}{-\alpha+1}\left[x^{-\alpha+1}\right]_1^t = \frac{1}{-\alpha+1}(t^{-\alpha+1}-1).$$

It follows that

$$\int_1^t \frac{1}{x^\alpha}dx \to \frac{1}{\alpha-1} \quad \text{for } t \to \infty.$$

Via Theorem 2.2.1 we conclude that

$$\sum_{n=1}^{\infty} \frac{1}{n^\alpha} \text{ is convergent if } \alpha > 1.$$

For $\alpha = 1$,

$$\int_1^t \frac{1}{x}dx = [\ln x]_1^t = \ln t \to \infty \quad \text{for } t \to \infty.$$

Again via Theorem 2.2.1 we conclude that

$$\sum_{n=1}^{\infty} \frac{1}{n} \text{ is divergent.}$$

Let us now find an estimate for the infinite sum $\sum_{n=1}^{\infty} \frac{1}{n^2}$, with an error of at most $\epsilon = 0.1$. To do so, we use Corollary 2.2.2, which shows that for any $N \in \mathbb{N}$,

$$\begin{aligned} \left| \sum_{n=1}^{\infty} \frac{1}{n^2} - \sum_{n=1}^{N} \frac{1}{n^2} \right| &\leq \int_{N+1}^{\infty} \frac{1}{x^2}dx + \frac{1}{(N+1)^2} \\ &= \frac{1}{N+1} + \frac{1}{(N+1)^2}. \end{aligned} \tag{2.7}$$

The term in (2.7) is smaller than 0.1 already for $N = 10$; this shows that with an error of maximally 0.1,

$$\sum_{n=1}^{\infty} \frac{1}{n^2} \approx \sum_{n=1}^{10} \frac{1}{n^2} = 1.55.$$

We shall in Example 3.6.2 prove that

$$\sum_{n=1}^{\infty} \frac{1}{n^2} = \frac{\pi^2}{6} \approx 1.645. \qquad \square$$

Corollary 2.2.2 only applies to series with positive terms. Another case, where an estimate for how well the finite partial sums approximate an infinite sum can be given, is for *alternating series;* by this, we mean a series which can be written in the form

$$\sum_{n=1}^{\infty}(-1)^{n-1}b_n = b_1 - b_2 + b_3 - \cdots + (-1)^{n-1}b_n + \cdots, \tag{2.8}$$

where

$$b_n \geq 0 \text{ for all } n \in \mathbb{N}, \text{ or } b_n \leq 0 \text{ for all } n \in \mathbb{N}.$$

The following criterion is useful in order to check convergence for an alternating series. It can also be used to find an estimate for how many terms we need to include in the partial sum in order to reach a certain approximation of the sum:

Proposition 2.2.4 *An alternating series for which the sequence* $\{b_n\}_{n=1}^{\infty}$ *decreases monotonically to zero, is convergent, and for all* $N \in \mathbb{N}$,

$$\left|\sum_{n=1}^{\infty}(-1)^{n-1}b_n - \sum_{n=1}^{N}(-1)^{n-1}b_n\right| \leq b_{N+1}.$$

Example 2.2.5 By Proposition 2.2.4, the series

$$\sum_{n=1}^{\infty}(-1)^{n-1}\frac{1}{n}$$

is convergent. According to Example 2.2.3, the series is not absolutely convergent. As already mentioned, this implies that for any given number $\tilde{S}$ we can reorder the infinite series so that we obtain a convergent series with sum equal to $\tilde{S}$. Furthermore, it can be reordered so we obtain a divergent series. □

Corollary 2.2.2 and Proposition 2.2.4 are efficient tools to approximate infinite sums, whenever they apply. For general series with real terms it could be tempting to try to reorder the series in such a way that these results could be applied; however, as we have seen, this is only allowed for unconditionally convergent series.

2.3 Geometric series

The next step is to consider series where the terms depend on a variable. One of the simplest examples is a *geometric series;* by this we mean a series of the form

$$\sum_{n=0}^{\infty} x^n = 1 + x + x^2 + \cdots + x^n + \cdots, \tag{2.9}$$

where $x \in \mathbb{R}$ (or more generally, $x \in \mathbb{C}$). When dealing with a specific value

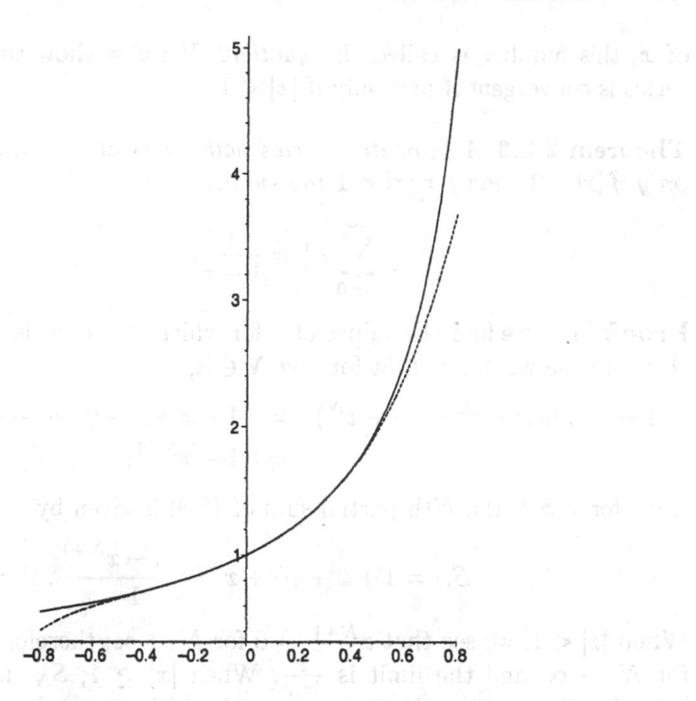

Figure 2.3.1 *The function $f(x) = \frac{1}{1-x}$ and the partial sum $S_5(x) = \sum_{n=0}^{5} x^n$ of the series representation in (2.10) (dotted).*

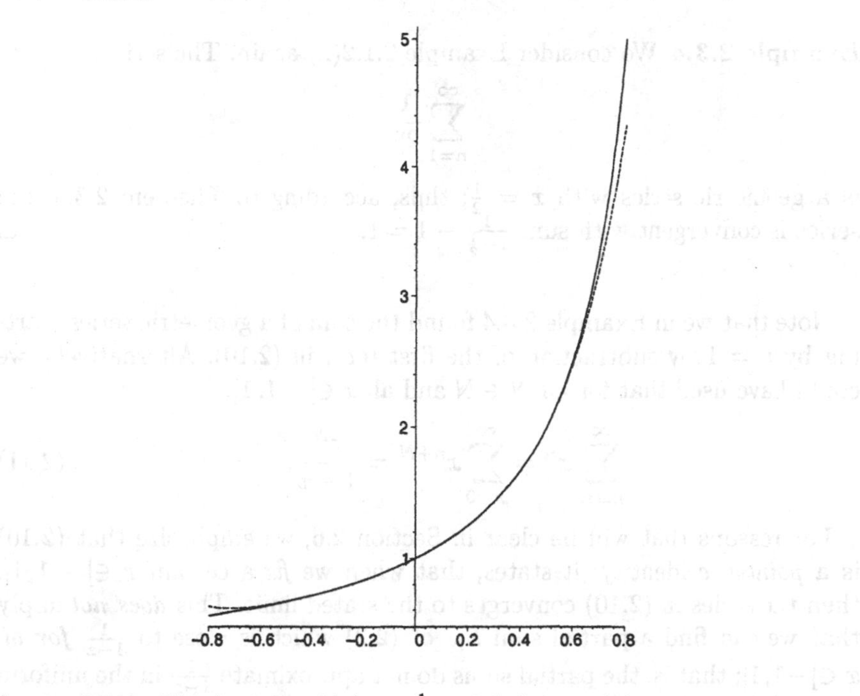

Figure 2.3.2 *The function $f(x) = \frac{1}{1-x}$ and the partial sum $S_8(x) = \sum_{n=0}^{8} x^n$ (dotted).*

of x, this number is called the *quotient*. We now show that a geometric series is convergent if and only if $|x| < 1$:

Theorem 2.3.3 *A geometric series with quotient x is convergent if and only if $|x| < 1$; and for $|x| < 1$ the sum is*

$$\sum_{n=0}^{\infty} x^n = \frac{1}{1-x}. \tag{2.10}$$

Proof: First we find the values of x for which the series is convergent. For this purpose we notice that for any $N \in \mathbb{N}$,

$$\begin{aligned}(1-x)(1+x+x^2+\cdots+x^N) &= 1-x+x-x^2+\cdots+x^N-x^{N+1}\\ &= 1-x^{N+1};\end{aligned}$$

thus, for $x \neq 1$, the Nth partial sum of (2.9) is given by

$$S_N = 1+x+\cdots+x^N = \frac{1-x^{N+1}}{1-x}.$$

When $|x| < 1$, we see that $x^{N+1} \to 0$ for $N \to \infty$; therefore S_N converges for $N \to \infty$ and the limit is $\frac{1}{1-x}$. When $|x| \geq 1$, S_N does not have a limit for $N \to \infty$ (nor in the case $x = 1$ which was excluded in the above discussion). □

Example 2.3.4 We consider Example 2.1.2(ii) again. The series

$$\sum_{n=1}^{\infty} \frac{1}{2^n}$$

is a geometric series with $x = \frac{1}{2}$; thus, according to Theorem 2.3.3, the series is convergent with sum $\frac{1}{1-\frac{1}{2}} - 1 = 1$. □

Note that we in Example 2.3.4 found the sum of a geometric series starting by $n = 1$ by subtraction of the first term in (2.10). Alternatively, we could have used that for all $N \in \mathbb{N}$ and all $x \in]-1,1[$,

$$\sum_{n=N}^{\infty} x^n = \sum_{n=0}^{\infty} x^{n+N} = \frac{x^N}{1-x}. \tag{2.11}$$

For reasons that will be clear in Section 2.6, we emphasize that (2.10) is a *pointwise* identity: it states, that when we *fix* a certain $x \in]-1,1[$, then the series in (2.10) converges to the stated limit. This *does not* imply that we can find a partial sum S_N of (2.9) which is close to $\frac{1}{1-x}$ *for all* $x \in]-1,1[$; that is, the partial sums do not approximate $\frac{1}{1-x}$ in the uniform sense of (1.2). See Figures 2.3.1–2.3.2; we return to the question of uniform approximation in Section 2.6.

2.4 Power series

We are now ready to connect the infinite series with the approximation theory discussed in Chapter 1. We begin with a reformulation of Theorem 1.3.6 using infinite series:

Theorem 2.4.1 *Let I be an interval and f an arbitrarily often differentiable function defined on I. Let $x_0 \in I$, and assume that there exists a constant $C > 0$ such that*

$$\left|f^{(n)}(x)\right| \leq C \quad \text{for all } n \in \mathbb{N} \quad \text{and all } x \in I.$$

Then

$$f(x) = \sum_{n=0}^{\infty} \frac{f^{(n)}(x_0)}{n!}(x - x_0)^n \quad \text{for all } x \in I. \tag{2.12}$$

Proof: The statement in Theorem 1.3.6 shows that the Nth partial sum of the infinite series in (2.12) converges to $f(x)$ when $N \to \infty$; thus, the result follows by the definition of an infinite sum. □

In the remainder of this chapter we will concentrate on the case where $0 \in I$ and we can choose $x_0 = 0$.

Example 2.4.2 Let us go back to Example 1.3.3, where we considered the function

$$f(x) = e^x, \ x \in]-1, 2[.$$

Since

$$f'(x) = f''(x) = \cdots = e^x,$$

we see that

$$|f^{(n)}(x)| \leq e^2 \text{ for all } n \in \mathbb{N}, \ x \in]-1, 2[. \tag{2.13}$$

Thus, according to Theorem 2.4.1,

$$f(x) = \sum_{n=0}^{\infty} \frac{f^{(n)}(0)}{n!} x^n, \ x \in]-1, 2[. \tag{2.14}$$

The particular interval $]-1, 2[$ does not play any role in (2.14): the arguments leading to this identity can be performed for any bounded interval containing 0. This implies that the identity holds pointwise for *any* $x \in \mathbb{R}$. Using that $f^{(n)}(0) = 1$ for all $n \in \mathbb{N}$, we arrive at

$$e^x = \sum_{n=0}^{\infty} \frac{x^n}{n!}, \ x \in \mathbb{R}. \tag{2.15}$$

Formula (2.15) shows that calculation of e^x can be reduced to repeated use of the operations addition, multiplication, and division. All calculators (and mathematical calculating programs like Maple, Mathematica or Matlab) are based only on these operations (together with subtraction) and all other computations are done by reduction to these. In fact, calculators evaluate expressions involving e^x using series expansions like the one in (2.15)! However, a mathematical program can not add infinitely many terms in finite time, so in practice it calculates

$$\sum_{n=0}^{N} \frac{x^n}{n!} \approx e^x$$

for a large value of N; see Example 1.3.7, where we saw that $N = 8$ was sufficient to approximate e^x with an error of at most 0.015 on the interval $[-1, 2]$. The commercial mathematical programs use way more terms, and the magnitude of the round-off error is considerably smaller; under "normal" conditions one will not notice the round-off error, but sometimes with long calculations the errors add up and become noticeable. □

Via the result in (2.15) we can now find an exact value for the integral considered in Example 1.3.8, in terms of an infinite series:

Example 2.4.3 In Example 1.3.8, we gave an estimate for the integral

$$\int_0^1 e^{-x^2}\, dx.$$

Now, by (2.15),

$$e^{-x^2} = \sum_{n=0}^{\infty} \frac{(-x^2)^n}{n!}, \quad x \in \mathbb{R}.$$

One can prove that it is allowed to integrate a power series over a bounded interval term-wise, see Corollary 2.6.8 ; in the present case, this shows that

$$\begin{aligned}
\int_0^1 e^{-x^2}\, dx &= \int_0^1 \left(\sum_{n=0}^{\infty} \frac{(-x^2)^n}{n!} \right) dx \\
&= \sum_{n=0}^{\infty} \frac{(-1)^n}{n!} \int_0^1 x^{2n}\, dx \\
&= \left[\sum_{n=0}^{\infty} (-1)^n \frac{x^{2n+1}}{(2n+1)n!} \right]_{x=0}^{1} \\
&= \sum_{n=0}^{\infty} (-1)^n \frac{1}{(2n+1)n!}. \qquad (2.16)
\end{aligned}$$

Thus, we have now obtained an exact expression for the integral in terms of an infinite series. In contrast, Example 1.3.8 gave an approximative value. We can obtain an arbitrarily precise value for the integral, simply by including sufficiently many terms in the partial sum of (2.16). We can even estimate how many terms we need to include in order to obtain a certain precision: according to Proposition 2.2.4,

$$\left|\sum_{n=0}^{\infty}(-1)^n\frac{1}{(2n+1)n!}-\sum_{n=0}^{N}(-1)^n\frac{1}{(2n+1)n!}\right|\leq\frac{1}{(2N+3)(N+1)!}.$$

This result shows that $N=8$ leads to a better approximation of the integral than suggested by Exercise 1.3; see Exercise 2.6. □

Example 2.4.4 Consider the function $f(x)=\sin x,\ x\in\mathbb{R}$. Then

$$\begin{aligned}
f'(x) &= \cos x = f^{(1+4n)}(x), \text{ and } f^{(1+4n)}(0)=1,\ n\in\mathbb{N},\\
f''(x) &= -\sin x = f^{(2+4n)}(x), \text{ and } f^{(2+4n)}(0)=0,\ n\in\mathbb{N},\\
f^{(3)}(x) &= -\cos x = f^{(3+4n)}(x), \text{ and } f^{(3+4n)}(0)=-1,\ n\in\mathbb{N},\\
f^{(4)}(x) &= \sin x = f^{(4n)}(x), \text{ and } f^{(4n)}(0)=0,\ n\in\mathbb{N}.
\end{aligned}$$

We see that $|f^{(n)}(x)|\leq 1$ for all $n\in\mathbb{N}$ and all $x\in\mathbb{R}$. By Theorem 2.4.1,

$$\begin{aligned}
\sin x &= \sum_{n=0}^{\infty}\frac{f^{(n)}(0)}{n!}x^n\\
&= x-\frac{x^3}{3!}+\frac{x^5}{5!}-\cdots\\
&= \sum_{n=0}^{\infty}\frac{(-1)^n}{(2n+1)!}x^{2n+1}. \qquad (2.17)
\end{aligned}$$

Our calculators find $\sin x$ using this expression. The partial sums of (2.17) converge very fast to the sine function, mainly because of the denominator $(2n+1)!$ in (2.17), which makes the coefficient in the power series decay quickly. Another reason for the fast convergence is the oscillatory behavior of the terms in the partial sums: the series (2.17) is an alternating series, and according to Proposition 2.2.4,

$$\left|\sin x-\sum_{n=0}^{N}\frac{(-1)^n}{(2n+1)!}x^{2n+1}\right|\leq\frac{|x|^{2N+3}}{(2N+3)!}. \qquad (2.18)$$

If we want the difference in (2.18) to be smaller than 0.01 for all $x\in[-5,5]$, it is enough to take $N\geq 7$; for $N=7$, the corresponding partial sum is a polynomial of degree 15.

It is actually surprising that the series representation (2.17) of the sine function is valid for all $x\in\mathbb{R}$. None of the functions that appear in the series representation are periodic, but nevertheless the infinite sum, that is, the sine function, is periodic with period 2π! See Figure 2.4.7, which shows the partial sum

$$S_5(x) = x - \frac{x^3}{3!} + \frac{x^5}{5!};$$

this is clearly not a 2π-periodic function. This illustrates that very contra-intuitive things might happen for infinite series; we will meet several other such instances later. □

The geometric series, as well as the representations (2.15) and (2.17) derived for the exponential function and the sine function, all have the form $\sum_{n=0}^{\infty} a_n x^n$ for some coefficients a_n. In general, a series of the form $\sum_{n=0}^{\infty} a_n x^n$ is called a *power series*, and a function which can be written in the form

$$f(x) = \sum_{n=0}^{\infty} a_n x^n$$

for some coefficients a_n, is said to have a *power series representation*.

Power series appear in many contexts, where one needs to approximate complicated functions. Unfortunately not all functions can be represented with help of a power series; we will come back to this on page 32.

Concerning convergence of power series, a very interesting result holds: basically, it says that a power series always is convergent on a certain symmetric interval around $x = 0$ and divergent outside the interval.

Theorem 2.4.5 *For every power series $\sum_{n=0}^{\infty} a_n x^n$ one of the following options holds:*

(i) The series only converges for $x = 0$.

(ii) The series is absolutely convergent for all $x \in \mathbb{R}$.

(iii) There exists a number $\rho > 0$ such that the series is absolutely convergent for $|x| < \rho$, and divergent for $|x| > \rho$.

We give a proof of this result (under an extra assumption) in Section A.4. The number ρ in (iii) is called *radius of convergence* of the power series. In the case (i) we put $\rho = 0$, and in (ii), $\rho = \infty$. Theorem 2.4.5 also holds for complex values of x, which explains the word "radius of convergence"; see Figure 2.4.8, which corresponds to the case $\rho = 1$.

Example 2.4.6 In order to find the radius of convergence for the power series $\sum_{n=0}^{\infty} \frac{x^n}{n!}$, we consider $x \neq 0$ and put $a_n = \frac{x^n}{n!}$. Then

$$\left|\frac{a_{n+1}}{a_n}\right| = \left|\frac{x^{n+1}}{(n+1)!}\,\frac{n!}{x^n}\right| = \frac{|x|}{n+1} \to 0 \quad \text{for } n \to \infty.$$

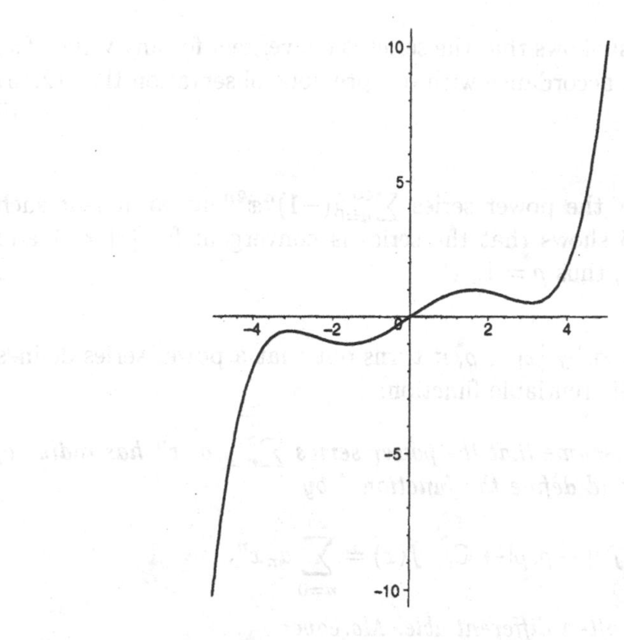

Figure 2.4.7 *The partial sum $S_5(x)$ for $f(x) = \sin x$.*

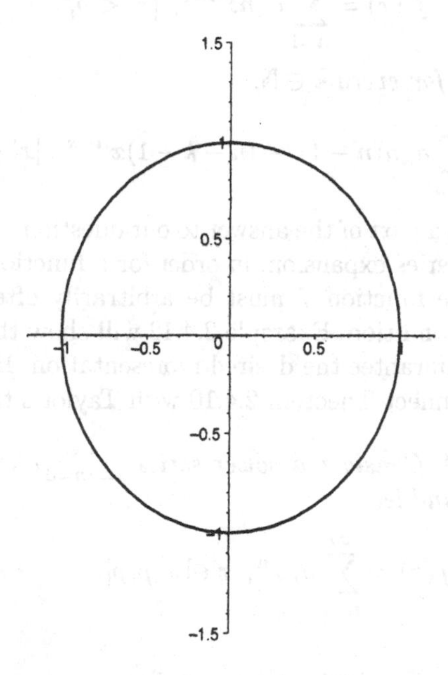

Figure 2.4.8 *Circle of convergence for the series $\sum_{n=0}^{\infty}(-1)^n x^{2n}$ in Example 2.4.9, with radius of convergence $\rho = 1$. The series is convergent for x-values on the part of the real axis which is inside the circle, and divergent outside the circle. Thinking about x as a complex variable and the plane $\mathbb{R}^2$ as $\mathbb{C}$, the series is convergent for all x inside the circle.*

Now the quotient test shows that the series is convergent for any value of x, so $\rho = \infty$. This is in accordance with our previous observation that (2.15) holds for all x. □

Example 2.4.9 For the power series $\sum_{n=0}^{\infty}(-1)^n x^{2n}$ an argument such as in Example 2.4.6 shows that the series is convergent for $|x| < 1$ and divergent for $|x| > 1$; thus $\rho = 1$. □

In the interval given by $|x| < \rho$, it turns out that a power series defines an infinitely often differentiable function:

Theorem 2.4.10 *Assume that the power series $\sum_{n=0}^{\infty} a_n x^n$ has radius of convergence $\rho > 0$, and define the function f by*

$$f :]-\rho, \rho[\to \mathbb{C}, \quad f(x) = \sum_{n=0}^{\infty} a_n x^n.$$

Then f is infinitely often differentiable. Moreover

$$f'(x) = \sum_{n=1}^{\infty} a_n n x^{n-1}, \ |x| < \rho,$$

and more generally, for every $k \in \mathbb{N}$,

$$f^{(k)}(x) = \sum_{n=k}^{\infty} a_n n(n-1)\cdots(n-k+1)x^{n-k}, \ |x| < \rho. \tag{2.19}$$

The theorem gives a part of the answer to our question about which functions have a power series expansion: in order for a function f to have such a representation, the function f must be arbitrarily often differentiable: this is a *necessary* condition. Example 2.4.12 will show that the condition is not sufficient to guarantee the desired representation. Before we present that example, we connect Theorem 2.4.10 with Taylor's theorem.

Proposition 2.4.11 *Consider a power series $\sum_{n=0}^{\infty} a_n x^n$ with radius of convergence $\rho > 0$, and let*

$$f(x) = \sum_{n=0}^{\infty} a_n x^n, \ x \in]-\rho, \rho[.$$

Then

$$a_n = \frac{f^{(n)}(0)}{n!}, \ n = 0, 1, \ldots, \tag{2.20}$$

i.e.,

$$f(x) = \sum_{n=0}^{\infty} \frac{f^{(n)}(0)}{n!} x^n, \ x \in]-\rho, \rho[.$$

Proof: First, $f(0) = a_0$. Letting $x = 0$ in Theorem 2.4.10 we get that

$$f'(0) = a_1, \; f''(0) = 2 \cdot 1 \cdot a_2,$$

and more generally, via (2.19),

$$f^{(k)}(0) = a_k \cdot k!, \; k \in \mathbb{N}.$$

□

Proposition 2.4.11 shows that if a function f has a power series representation with some arbitrary coefficients a_n, then it is *necessarily* the representation we saw in Theorem 2.4.1. In particular, the power series representation is *unique*, i.e., only one choice of the coefficients a_n is possible. In Chapter 5 we will see examples of other types of series representations, where there are many possible ways to choose the coefficients.

As promised, we now show that not every arbitrarily often differentiable function $f : \mathbb{R} \to \mathbb{R}$ has a power series representation. We do so by example, i.e., we exhibit a concrete function without such a representation:

Example 2.4.12 Consider a function $f : \mathbb{R} \to \mathbb{R}$ with the following properties:

(i) $f(x) = 1$ for all $x \in [-1, 1]$.

(ii) f is arbitrarily often differentiable.

Assume that f has a power series representation,

$$f(x) = \sum_{n=0}^{\infty} a_n x^n \text{ for all } x \in \mathbb{R}. \tag{2.21}$$

Proposition 2.4.11 shows us that

$$a_n = \frac{f^{(n)}(0)}{n!} = \begin{cases} 1 \text{ if } n = 0, \\ 0 \text{ otherwise.} \end{cases}$$

Inserting this in (2.21) shows that

$$f(x) = \sum_{n=0}^{\infty} a_n x^n = 1, \; \forall x \in \mathbb{R}.$$

It follows that unless f is a constant function, f does not have a power series representation that holds for all $x \in \mathbb{R}$. So all we have to do is to exhibit a concrete function $f : \mathbb{R} \to \mathbb{R}$, which is not constant, but satisfies (i) and (ii). One choice is the function

$$f(x) = \begin{cases} 1 \text{ for } |x| \leq 1, \\ 1 - e^{\frac{-1}{x^2-1}} \text{ for } |x| > 1. \end{cases} \tag{2.22}$$

This function is shown in Figure 2.5.1. At a first glance, the function does not appear to be differentiable at $x = 1$! However, that this is indeed the case seems more reasonable from Figure 2.5.2, which shows the function in a small neighborhood of $x = 1$.

A mathematical proof of the fact that the function is arbitrarily often differentiable falls outside the scope of this book. □

The result in Example 2.4.12 can be formulated differently. In fact the example exhibits a function f for which the Taylor series $\sum_{n=0}^{\infty} \frac{f^{(n)}(0)}{n!} x^n$ converges for all $x \in \mathbb{R}$, but with a sum which does not equal $f(x)$ for all $x \in \mathbb{R}$.

We notice that it is known exactly which functions can be represented via a power series: it is the class of analytic functions (i.e., the functions that are complex differentiable). More details on this can be found in any standard textbook on complex analysis, e.g. [7].

2.5 General infinite sums of functions

As we have seen, a power series consists of infinite sums of terms $a_n x^n$, $n \in \mathbb{N}$. Since not all functions have a representation of this type, it is natural also to consider infinite sums of other types of "simple" functions. More generally, we will consider a family of functions $f_0, f_1, f_2, \dots$ with the *same* domain of definition I, and attempt to define the function

$$f(x) = f_0(x) + f_1(x) + f_2(x) + \cdots + f_n(x) + \cdots = \sum_{n=0}^{\infty} f_n(x). \quad (2.23)$$

Here it is of course important to know for which values of x the expression for $f(x)$ makes sense: this is the case exactly for the $x \in I$ for which *the series of numbers* $\sum_{n=0}^{\infty} f_n(x)$ is convergent. The function f is called *the sum function* associated to $f_0, f_1, \dots$. As before we will call the expression

$$S_N(x) = f_0(x) + f_1(x) + \cdots + f_N(x)$$

the Nth partial sum of $\sum_{n=0}^{\infty} f_n(x)$; it is now a function depending on x.

In practice, one can only deal with a representation of the type (2.23) if the functions f_n have some kind of common structure. In the next chapters we will consider infinite series where the functions f_n are trigonometric functions or elements from a wavelet system; both types of systems satisfy this requirement.

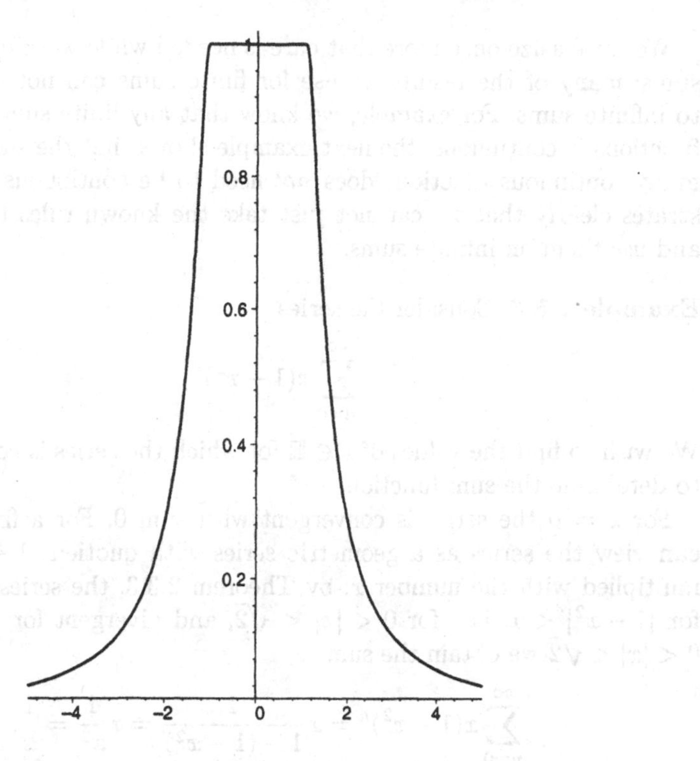

Figure 2.5.1 *The function in* (2.22), *shown on the interval* $[-5, 5]$.

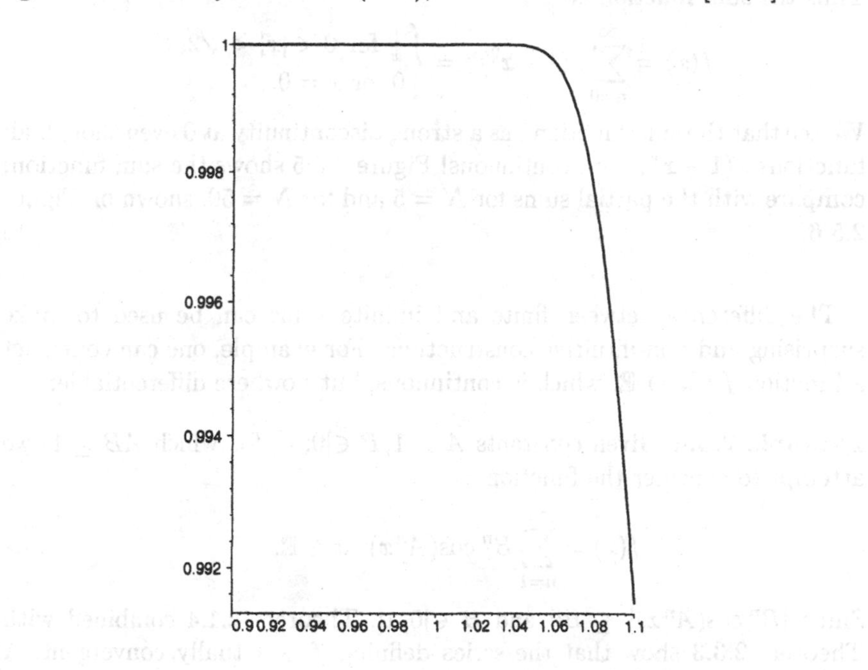

Figure 2.5.2 *The function in* (2.22), *shown on the interval* $[0.9, 1.1]$. *Notice the unit on the second axis!*

We emphasize once more that care is needed while working with infinite sums: many of the results we use for finite sums can not be generalized to infinite sums. For example, we know that any finite sum of continuous functions is continuous; the next example shows that the sum of infinitely many continuous functions does *not* need to be continuous. This demonstrates clearly that we can not just take the known rules for finite sums and use them on infinite sums.

Example 2.5.3 Consider the series

$$\sum_{n=0}^{\infty} x(1-x^2)^n.$$

We wish to find the values of $x \in \mathbb{R}$ for which the series is convergent, and to determine the sum function.

For $x = 0$ the series is convergent with sum 0. For a fixed $x \neq 0$ we can view the series as a geometric series with quotient $1 - x^2$, which is multiplied with the number x; by Theorem 2.3.3, the series is convergent for $|1 - x^2| < 1$, i.e., for $0 < |x| < \sqrt{2}$, and divergent for $|x| \geq \sqrt{2}$. For $0 < |x| < \sqrt{2}$ we obtain the sum

$$\sum_{n=0}^{\infty} x(1-x^2)^n = x\frac{1}{1-(1-x^2)} = x\frac{1}{x^2} = \frac{1}{x}.$$

Thus the sum function is

$$f(x) = \sum_{n=0}^{\infty} x(1-x^2)^n = \begin{cases} \frac{1}{x} \text{ for } 0 < |x| < \sqrt{2}, \\ 0 \text{ for } x = 0. \end{cases}$$

We see that the sum function has a strong discontinuity at 0 even though all functions $x(1-x^x)^n$ are continuous! Figure 2.5.5 shows the sum function; compare with the partial sums for $N = 5$ and for $N = 50$, shown on Figure 2.5.6. □

The difference between finite and infinite sums can be used to make surprising and non-intuitive constructions. For example, one can construct a function $f : \mathbb{R} \to \mathbb{R}$, which is continuous, but nowhere differentiable:

Example 2.5.4 Given constants $A > 1, B \in]0,1[$, for which $AB \geq 1$, we attempt to consider the function

$$f(x) = \sum_{n=1}^{\infty} B^n \cos(A^n x), \ x \in \mathbb{R}.$$

Since $|B^n \cos(A^n x)| \leq B^n$ and $B \in]0,1[$, Theorem 2.1.4 combined with Theorem 2.3.3 show that the series defining f is actually convergent. A result stated formally in the next section, Theorem 2.6.5, tells us that f is continuous.

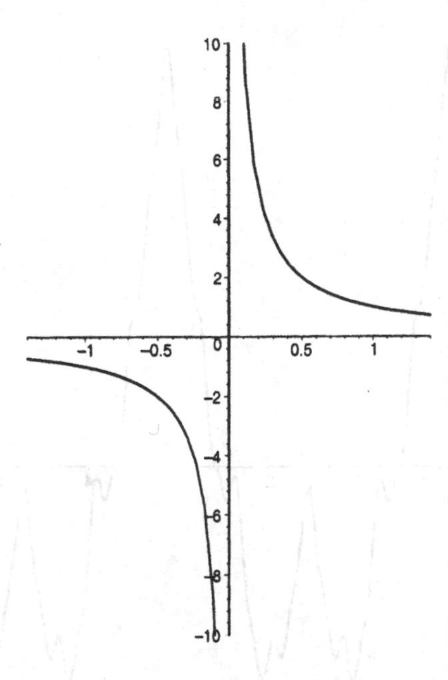

Figure 2.5.5 *The function* $f(x) = \sum_{n=0}^{\infty} x(1-x^2)^n, \ x \in]-\sqrt{2}, \sqrt{2}[$.

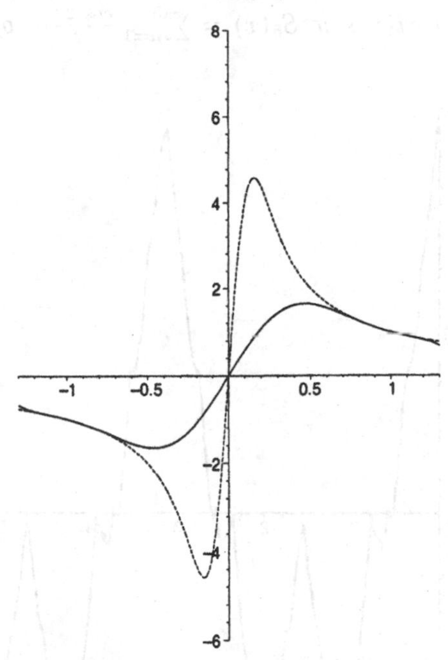

Figure 2.5.6 *The partial sums* $S_5(x) = \sum_{n=0}^{5} x(1-x^2)^n$ *(unbroken line) and* $S_{50}(x) = \sum_{n=0}^{50} x(1-x^2)^n$ *(dotted). The more terms we include in the partial sum, the more the graph looks like the graph of the sum function in Figure 2.5.5. However, all partial sums "turn off" near the point* $(0,0)$, *where all the graphs go through.*

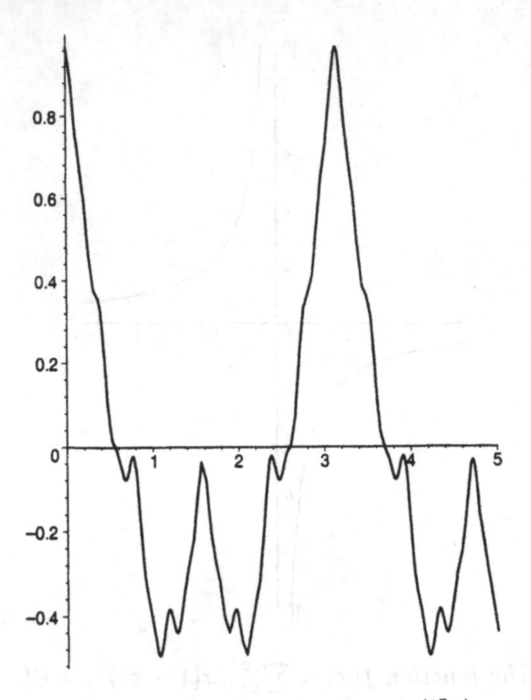

Figure 2.5.7 *The partial sum* $S_5(x) = \sum_{n=1}^{5} \frac{\cos(2^n x)}{2^n}$ *of the function in* (2.24).

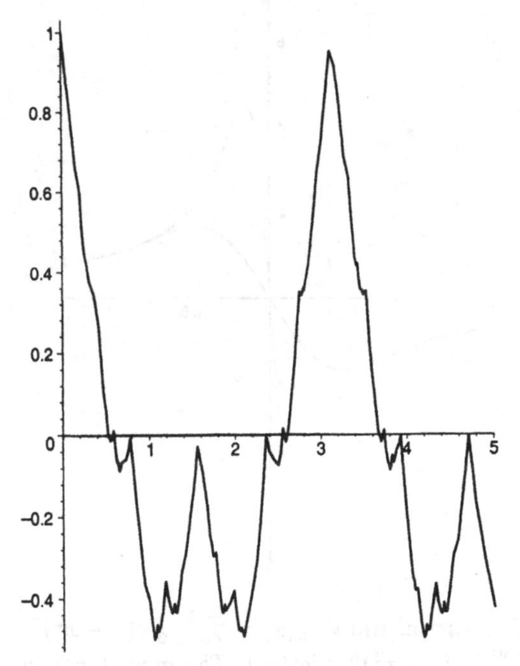

Figure 2.5.8 *The partial sum* $S_{50}(x) = \sum_{n=1}^{50} \frac{\cos(2^n x)}{2^n}$ *of the function in* (2.24).

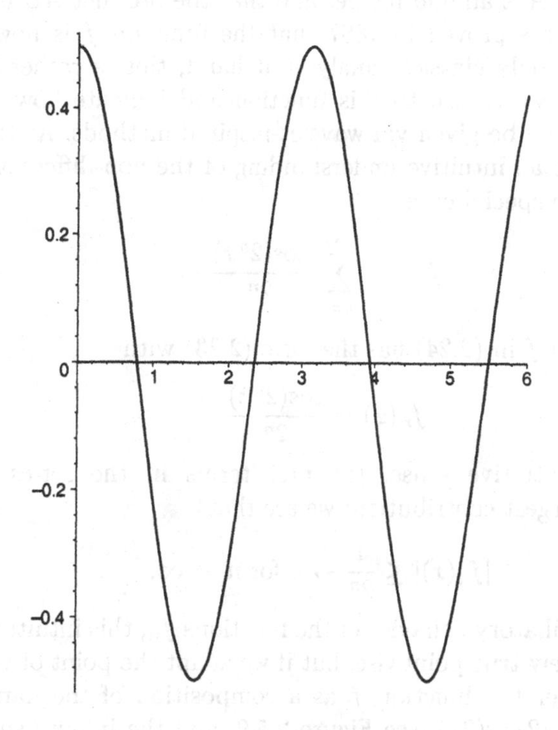

Figure 2.5.9 *The function* $f_1(x) = \frac{1}{2}\cos(2x)$.

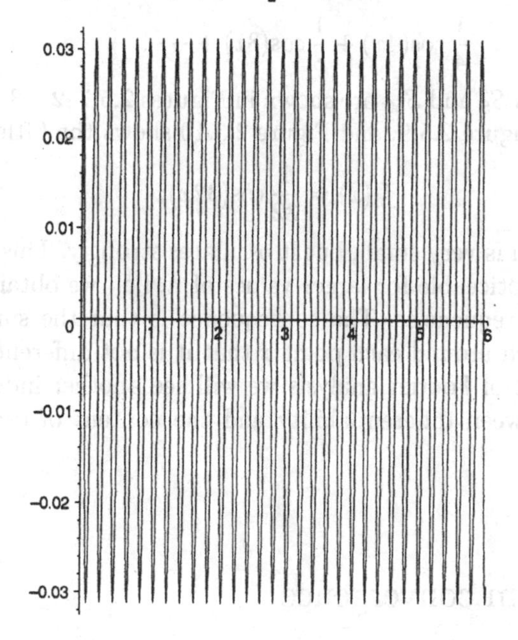

Figure 2.5.10 *The function* $f_5(x) = \frac{1}{32}\cos(32x)$. *Observe the units on the axes!*

Assuming that A is an odd integer and that the product AB is sufficiently large, Weierstrass proved in 1887 that the function f is nowhere differentiable. With only classical analysis at hand, this is rather involved; in Example 5.6.4 we return to this function and indicate how a short and elegant proof can be given via wavelet-inspired methods. At this moment, we only aim at an intuitive understanding of the non-differentiability. Let us consider the special case

$$f(x) = \sum_{n=1}^{\infty} \frac{\cos(2^n x)}{2^n}. \tag{2.24}$$

The function f in (2.24) has the form (2.23) with

$$f_n(x) = \frac{\cos(2^n x)}{2^n}.$$

In a rather intuitive sense, the first terms in the series deliver the numerically largest contribution: we see that

$$|f_n(x)| \leq \frac{1}{2^n} \to 0 \text{ for } n \to \infty.$$

Due to the oscillatory behavior of the functions f_n, this intuitive statement is not completely true pointwise, but if we adapt the point of view anyway, we can consider the function f as a composition of the harmonic function $f_1(x) = 1/2\cos(2x)$, see Figure 2.5.9, and the infinite sum of higher harmonics

$$\frac{1}{4}\cos(4x) + \frac{1}{8}\cos(8x) + \cdots .$$

The partial sums S_5 and S_{50} are shown in Figures 2.5.7–2.5.8. The function f_1 is shown in Figure 2.5.9, and Figure 2.5.10 shows the fifth term

$$f_5(x) = \frac{1}{32}\cos(32x);$$

already this term is very small, but it oscillates strongly. This phenomenon becomes more noticeable for larger values of n, i.e., we obtain oscillations with very high frequencies. The consequence is that the sum function f oscillates so much around each point x that it is not differentiable.

In the context of Fourier analysis we will see another indication of the relationship between differentiability and the content of oscillations; see Section 3.7. □

2.6 Uniform convergence

So far, our definition of convergence of an infinite series of functions has been ***pointwise:*** in fact, (2.23) merely means that if we fix an arbitrary x in the domain of f, the partial sums of the series converge to $f(x)$. However, as we will explain now, this type of convergence is not sufficient in all situations.

As already mentioned, the motivation behind (2.23) is to write a complicated signal f as an infinite sum of simpler functions. However, since a computer can only deal with finite sums, for applications of this result in practice we always have to replace the infinite sum in (2.23) by a *finite sum*, i.e., by a certain partial sum. In other words, we need to be able to find $N \in \mathbb{N}$ such that

$$|f(x) - S_N(x)|$$

is sufficiently small, ***simultaneously for all x in the considered interval.*** That is, in the terminology used in Section 1.1, we need that the partial sums deliver approximations of f in the uniform sense. As the following example demonstrates, this is not always the case.

Example 2.6.1 Consider the function

$$f(x) = \frac{1}{1-x}, \quad x \in]-1,1[.$$

Then, as we saw in Theorem 2.3.3,

$$f(x) = \sum_{n=0}^{\infty} x^n, \tag{2.25}$$

understood in the sense that the partial sums $S_N(x)$ of the right-hand side of (2.25) converges to $f(x)$ for any $x \in]-1,1[$. However, if we fix any $N \in \mathbb{N}$, we have

$$\begin{aligned} \left| f(x) - \sum_{n=0}^{N} x^n \right| &= \left| \sum_{n=N+1}^{\infty} x^n \right| \\ &= \frac{|x|^{N+1}}{1-x}. \end{aligned}$$

No matter how large we choose $N \in \mathbb{N}$, there exists an $x \in]-1,1[$ for which this difference is as large as we would like to ask for (see Figures 2.3.1–2.3.2). Thus, we are not able to obtain a good approximation simultaneously for all $x \in]-1,1[$; this is a serious shortcoming for practical applications.

Note that the above problem disappears if we only consider the function f on any interval of the type $[-r,r]$ for some $r \in]0,1[$; in this case, for any $x \in [-r,r]$ we have

$$\begin{aligned} \left| f(x) - \sum_{n=0}^{N} x^n \right| &= \frac{|x|^{N+1}}{1-x} \\ &\leq \frac{r^{N+1}}{1-r}. \end{aligned}$$

Since

$$\frac{r^{N+1}}{1-r} \to 0 \text{ as } N \to \infty,$$

we can thus make $\left|f(x) - \sum_{n=0}^{N} x^n\right|$ as small as we want, simultaneously for all $x \in [-r, r]$, by choosing $N \in \mathbb{N}$ sufficiently large. □

Example 2.6.1 suggests that we introduce the following type of convergence.

Definition 2.6.2 *Given functions $f_1, f_2, \ldots, f_n, \ldots$ defined on an interval I, assume that the function*

$$f(x) = \sum_{n=1}^{\infty} f_n(x), \ x \in I,$$

is well defined. Then we say that $\sum_{n=1}^{\infty} f_n$ converges uniformly to f on the interval I if for each $\epsilon > 0$ we can find an $N_0 \in \mathbb{N}$ such that

$$\left|f(x) - \sum_{n=1}^{N} f_n(x)\right| \leq \epsilon \text{ for all } x \in I \text{ and all } N \geq N_0.$$

Note that an equivalent definition would be to say that $\sum_{n=1}^{\infty} f_n(x)$ converges uniformly to f on the interval I if for each $\epsilon > 0$ we can find $N_0 \in \mathbb{N}$ such that

$$\sup_{x \in I} \left|f(x) - \sum_{n=1}^{N} f_n(x)\right| \leq \epsilon \text{ for all } N \geq N_0.$$

Let us return to Example 2.6.1:

Example 2.6.3 Formulated in terms of uniform convergence, our result in Example 2.6.1 shows that

- $\sum_{n=0}^{\infty} x^n$ does not converge uniformly to $f(x) = \frac{1}{1-x}$ on $I =]-1, 1[$;
- $\sum_{n=0}^{\infty} x^n$ converges uniformly to $f(x) = \frac{1}{1-x}$ on any interval $I = [-r, r]$, $r \in]0, 1[$.

It turns out that Example 2.6.3 is typical for power series: in general, nothing guarantees that a power series with radius of convergence ρ converges uniformly on $]-\rho, \rho[$. On the other hand we always obtain uniform convergence if we shrink the interval slightly:

Proposition 2.6.4 *Let $\sum_{n=0}^{\infty} a_n x^n$ be a power series with radius of convergence $\rho > 0$. Then $\sum_{n=0}^{\infty} a_n x^n$ is uniformly convergent on any interval of the form $[-r, r]$, where $r \in]0, \rho[$.*

Proof: Given any $x \in [-r, r]$,

$$\left|\sum_{n=0}^{\infty} a_n x^n - \sum_{n=0}^{N} a_n x^n\right| = \left|\sum_{n=N+1}^{\infty} a_n x^n\right| \leq \sum_{n=N+1}^{\infty} |a_n x^n| \leq \sum_{n=N+1}^{\infty} |a_n r^n|;$$

thus

$$\sup_{x\in[-r,r]} \left|\sum_{n=0}^{\infty} a_n x^n - \sum_{n=0}^{N} a_n x^n\right| \leq \sum_{n=N+1}^{\infty} |a_n r^n|. \tag{2.26}$$

By the fact that $\sum_{n=1}^{\infty} a_n r^n$ is absolutely convergent, the quantity on the right-hand side of (2.26) goes to zero as $N \to \infty$. □

We end this section with a few important results concerning continuity and differentiability of infinite series of functions; the first result below gives conditions for the associated sum function being well defined and continuous.

Theorem 2.6.5 *Assume that the functions $f_1, f_2, \ldots$ are defined and continuous on an interval I, and that there exist positive constants $k_1, k_2, \ldots$ such that*

(i) $|f_n(x)| \leq k_n, \ \forall x \in I, \ n \in \mathbb{N}$;

(ii) $\sum_{n=1}^{\infty} k_n$ is convergent.

Then the infinite series $\sum_{n=1}^{\infty} f_n(x)$ converges uniformly; and the function

$$f(x) = \sum_{n=1}^{\infty} f_n(x), \ x \in I$$

is continuous.

A series $\sum_{n=1}^{\infty} k_n$ satisfying the condition (i) in Theorem 2.6.5 is called a *majorant series* for $\sum_{n=1}^{\infty} f_n(x)$. Theorem 2.6.5 is known in the literature under the name *Weierstrass' M-test.*

If we want the sum function to be differentiable, we need slightly different conditions, stated below; on the theoretical level, the result can be used to prove Theorem 2.4.10.

Theorem 2.6.6 *Assume that the functions $f_1, f_2, \ldots$ are defined and differentiable with a continuous derivative on the interval I, and that the* *function* $f(x) = \sum_{n=1}^{\infty} f_n(x)$ *is well defined on* I. *Assume also that there exist positive constants* $k_1, k_2, \ldots$ *such that*

(i) $|f_n'(x)| \leq k_n, \ \forall x \in I, \ n \in \mathbb{N}$;

(ii) $\sum_{n=1}^{\infty} k_n$ *is convergent.*

Then f *is differentiable on* I, *and*

$$f'(x) = \sum_{n=1}^{\infty} f_n'(x).$$

Finally, we mention that uniform convergence of an infinite series consisting of integrable functions allows us to integrate term-wise:

Proposition 2.6.7 *Assume that the functions* $f_1, f_2, \ldots$ *are continuous on the interval* I *and that* $\sum_{n=1}^{\infty} f_n(x)$ *is uniformly convergent. Then, the function* $x \mapsto \sum_{n=1}^{\infty} f_n(x)$ *is continuous on* I; *furthermore, for any* $a, b \in I$,

$$\int_a^b \sum_{n=1}^{\infty} f_n(x)dx = \sum_{n=1}^{\infty} \int_a^b f_n(x)dx.$$

Proposition 2.6.7 has an important consequence for integration of power series:

Corollary 2.6.8 *Let* $\sum_{n=0}^{\infty} a_n x^n$ *be a power series with radius of convergence* $\rho > 0$. *Then, for any* $b \in]-\rho, \rho[$,

$$\int_0^b \sum_{n=0}^{\infty} a_n x^n dx = \sum_{n=0}^{\infty} \frac{a_n}{n+1} b^{n+1}.$$

2.7 Signal transmission

Modern technology often requires that information can be sent from one place to another; one speaks about *signal transmission*. It occurs, e.g., in connection with wireless communication, the internet, computer graphics, or transfer of data from CD-ROM to computer.

All types of signal transmission are based on transmission of a series of numbers. The first step is to convert the given information (called the *signal*) to a series of numbers, and this is where the question of having a series representation comes in: if we know that a signal is given as a function f which has a power series representation

$$f(x) = \sum_{n=0}^{\infty} a_n x^n, \tag{2.27}$$

then all information about the function f is stored in the coefficients $\{a_n\}_{n=0}^{\infty}$. In other words: if we know the coefficients $\{a_n\}_{n=0}^{\infty}$, then we can immediately use (2.27) to find out which signal we are dealing with.

Let us explain how this can be used in signal transmission. Assume for example that a sender $\mathcal{S}$ wishes to send a picture to a receiver $\mathcal{R}$. For simplification, we assume that the picture is given as a graph of a function f, and that f can be represented as a power series. Then the transmission can be done in the following way:

- $\mathcal{S}$ finds the coefficients $a_0, a_1, \dots$ such that $f(x) = \sum_{n=0}^{\infty} a_n x^n$;
- $\mathcal{S}$ sends the coefficients $a_0, a_1, \dots$;
- $\mathcal{R}$ receives the coefficients $a_0, a_1, \dots$;
- $\mathcal{R}$ reconstructs the signal by multiplying the coefficients a_n by x^n and forming the infinite series $f(x) = \sum_{n=0}^{\infty} a_n x^n$.

In practice there are a few more steps not described here. For example, $\mathcal{S}$ can not send an infinite sequence of numbers $a_0, a_1, \dots$: it is possible to send only a finite sequence of numbers $a_0, a_1, \dots, a_N$. The choice of the numbers to send needs to be done with care: we want that the signal we actually transmit, i.e.,

$$f^{\sharp}(x) = a_0 + a_1 x + \cdots + a_N x^N = \sum_{n=0}^{N} a_n x^n,$$

looks and behaves like the original signal

$$f(x) = a_0 + a_1 x + \cdots + a_N x^N + \cdots = \sum_{n=0}^{\infty} a_n x^n.$$

This usually forces N to be large. Another issue is that $\mathcal{S}$ can only send the numbers $a_0, \dots, a_N$ with a certain precision, for example, 50 digits. So in reality $\mathcal{S}$ will send some numbers

$$\tilde{a}_0 \simeq a_0, \ \ \tilde{a}_1 \simeq a_1, \ \ \tilde{a}_2 \simeq a_2, \ \ \dots, \tilde{a}_N \simeq a_N,$$

and $\mathcal{R}$ receives coefficients corresponding to the signal

$$\tilde{f}(x) = \sum_{n=0}^{N} \tilde{a}_n x^n.$$

The fact that the numbers a_n can only be sent with a certain precision is well known in engineering; one speaks about *quantification.*

The process above needs to be made such that

$$\tilde{f}(x) \simeq f(x)$$

— it does not help if $\mathcal{S}$ sends a picture of an elephant and $\mathcal{R}$ receives something that looks like a giraffe! Sometimes it is unsuitable just to send the *first* $N+1$ coefficients $a_0, \ldots, a_N$: it will be much more reasonable to send the $N+1$ most important coefficients, which usually means the (numerically) *largest* coefficients. In this case one also has to send information about which powers x^n the coefficients that are sent belong to; we speak about *coding* of the information.

The idea of sending the $N+1$ largest coefficients instead of just the first $N+1$ coefficients can be seen as the first step towards *best* N*-term approximation*, a subject to which we return in Section 3.8 and Section 5.7. Observe that in the technical sense, to send the largest coefficients first usually corresponds to *reordering* the terms in the given infinite series; as we have seen in Section 2.1 this can lead to further complications if the series does not converge unconditionally, so it has to be done with care.

We also note that to send only a finite number of numbers $a_0, a_1, \ldots, a_N$, rather than the full sequence, corresponds to a *compression*, in the sense that a part of the information determining the signal is thrown away. In the situation discussed here, this type of compression can not be avoided: we need to have a finite sequence in order to be able to transmit. Usually the word "compression" appears in a slightly different sense, namely, that we want the information given in a finite sequence to be represented efficiently using a shorter sequence. This is a very important issue when dealing with large data sets; we will return to it in Section 4.2.

Example 2.7.1 Assume that $\mathcal{S}$ wishes to send the function

$$f(x) = \sin x, \; x \in [0,3]$$

to $\mathcal{R}$. In Example 2.4.4 we showed that

$$\sin x = \sum_{n=1}^{\infty} (-1)^{n+1} \frac{x^{2n+1}}{(2n+1)!} = 0 + x + 0 \cdot x^2 - \frac{1}{6}x^3 + 0 \cdot x^4 + \frac{1}{120}x^5 + \cdots .$$

Since $\frac{1}{6} \approx 0.1666$, $\frac{1}{120} \approx 0.0083$, $\mathcal{S}$ can choose to send the numbers

$$0; 1; \; 0; \; -0.1666; \; 0; \; 0.0083,$$

according to which $\mathcal{R}$ will reconstruct the function as

$$\tilde{f}(x) = x - 0.1666x^3 + 0.0083x^5 \approx f(x).$$

Figure 2.7.2 shows that the reconstructed signal is close to the original signal when $x \in [0,2]$, but then the approximation starts to deviate from the signal. A better reconstruction can be reached if we also send the coefficient belonging to x^7, namely $-\frac{1}{7!} \simeq -0.000198$; then

$$\tilde{f}(x) = x - 0.1666x^3 + 0.0083x^5 - 0.000198x^7,$$

see Figure 2.7.3. □

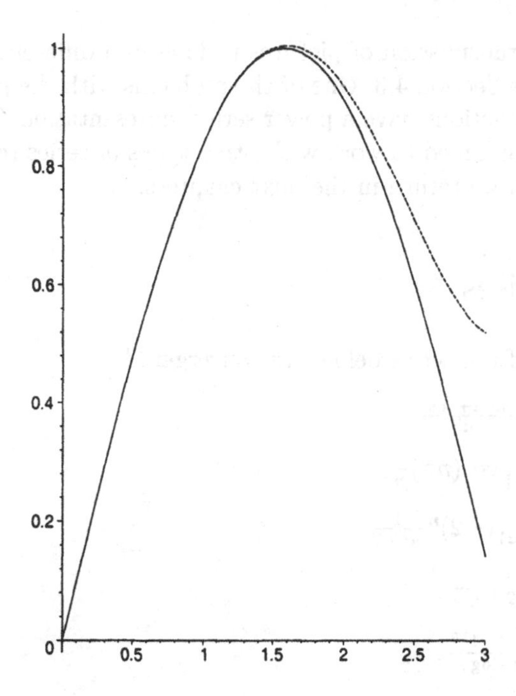

Figure 2.7.2 *The functions* $f(x) = \sin x$ *(unbroken line) and* $\tilde{f}(x) = x - 0.1666x^3 + 0.0083x^5$.

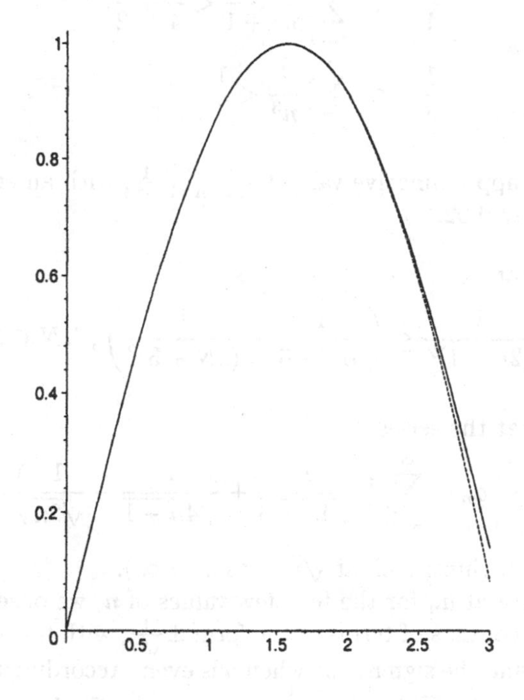

Figure 2.7.3 *The functions* $f(x) = \sin x$ *(unbroken line) and* $\tilde{f}(x) = x - 0.1666x^3 + 0.0083x^5 - 0.000198x^7$.

In practice, transmission of pictures is done in a different way than described here, see Section 4.3. One of the problems with the principle above is that not all functions have a power series representation. Thus, in many situations we are forced to work with other types of series representations, a topic to which we return in the next chapters.

2.8 Exercises

2.1 Which of the series below are convergent?

(i) $\sum_{n=1}^{\infty} \frac{\sin n + \cos n}{n^2}$

(ii) $\sum_{n=1}^{\infty} \cos(n\pi)\frac{2}{n+5}$

(iii) $\sum_{n=1}^{\infty} (-2)^n \frac{1}{n^2+7}$

(iv) $\sum_{n=2}^{\infty} \frac{1}{\log n}$

(v) $\sum_{n=2}^{\infty} \frac{(-1)^n}{\log n}$.

2.2 Prove the inequalities below via the integral test:

$$\frac{\pi}{4} < \sum_{n=1}^{\infty} \frac{1}{n^2+1} < \frac{\pi}{4} + \frac{1}{2},$$

$$\frac{1}{8} < \sum_{n=2}^{\infty} \frac{1}{n^3} < \frac{1}{4}.$$

2.3 Find an approximative value for $\sum_{n=1}^{\infty} \frac{1}{n^4}$, with an error of maximally 0.02.

2.4 Prove that

$$\sum_{n=N+1}^{\infty} \frac{1}{(2n+1)^2} \leq \left(\frac{1}{4N+6} + \frac{1}{(2N+3)^2} \right), \ \forall N \in \mathbb{N}.$$

2.5 Prove that the series

$$\sum_{n=1}^{\infty} a_n = \sum_{n=1}^{\infty} \left(\frac{1}{\sqrt{4n-3}} + \frac{1}{\sqrt{4n-1}} - \frac{1}{\sqrt{2n}} \right)$$

is divergent (hint: look at $\sqrt{n}a_n$ as $n \to \infty$).

Looking at a_n for the first few values of n, we observe that $\sum_{n=1}^{\infty} a_n$ consists of terms of the form $\pm\frac{1}{\sqrt{n}}$, with the sign plus when n is odd, and the sign minus when n is even. According to Proposition 2.2.4, the series $\sum_{n=1}^{\infty} \frac{(-1)^{n-1}}{\sqrt{n}}$ is convergent. Explain why these results do not contradict each other.

2.6 Compare the estimates for $\int_0^1 e^{-x^2}\,dx$ in

- Example 2.4.3, using the partial sums of (2.16) corresponding to $N = 5$ and $N = 8$;
- Exercise 1.3.

2.7 Find the radius of convergence for the power series

$$\sum_{n=1}^{\infty} nx^n,$$

and find an expression for the sum in terms of the standard functions.

2.8 Find the radius of convergence for the power series

$$\sum_{n=0}^{\infty} \frac{n+2}{n+1} x^{n+1},$$

and find an expression for the sum in terms of the standard functions.

2.9 (i) Find the radius of convergence for the power series

$$\sum\nolimits_{n=1}^{\infty} \frac{n+1}{n \cdot n!} x^n.$$

Let $f(x)$ denote the sum of this series.

(ii) Find a power series representation for the function

$$h(x) = \int_0^x t f'(t)dt, \; x \in \mathbb{R}.$$

(iii) Use the power series for e^x to find an expression for $h(x)$ in terms of the standard functions.

2.10 Prove that

$$f(x) := \sum\nolimits_{n=1}^{\infty} \frac{3\cos nx - \sin^2 nx}{3^n - 1}, \; x \in \mathbb{R},$$

defines a continuous function.

2.11 (i) Prove that the function

$$f(x) := \sum_{n=0}^{\infty} \frac{1}{3^n} \cos(2^n x), \; x \in \mathbb{R},$$

is well defined.

(ii) Prove that f is continuous.

(iii) Prove that f is differentiable, and find $f'(0)$.

2.12 Find the values of $x \in \mathbb{R}$ for which

$$\sum_{n=0}^{\infty} \frac{1}{(2+x^2)^n}$$

is convergent, and express the sum via the standard functions.

2.13 Prove that

$$\sum_{n=k+1}^{\infty} \frac{1}{(2n-1)^2} \leq \frac{1}{4k+2} + \frac{1}{(2k+1)^2}, \quad \forall k \in \mathbb{N}.$$

2.14 Find a polynomial $P(x)$ such that

$$|\sin x - P(x)| \leq 0.05 \text{ for all } x \in [0,3].$$

2.15 Assume that we want to transmit a function having a power series expansion. Explain why uniform convergence of the power series is relevant in this context.

3 Fourier Analysis

So far, we have mainly been dealing with power series representations, although we already saw the definition of an infinite series of more general functions. Unfortunately, only a relatively limited class of functions has a power series expansion, so often we need to seek other tools to represent functions.

Fourier series and the Fourier transform are such tools. Fourier analysis is a large and classical area of mathematics, dealing with representations of functions on $\mathbb{R}$ via trigonometric functions; as we will see, periodic functions have series expansions in terms of cosine functions and sine functions, while aperiodic functions f have expansions in terms of integrals involving trigonometric functions.

The sections below are of increasing complexity. Section 3.1 and Section 3.2, presenting Fourier series for periodic functions and their approximation properties, are understandable with the background obtained by reading Chapter 2; the same is the case with Section 3.6, dealing with Parseval's theorem. Section 3.3 motivates Fourier series from the perspective of signal analysis. Section 3.4 relates Fourier series with expansions in terms of orthonormal bases in vector spaces, and can be understood with prior knowledge of vector spaces with inner products. The rest of the chapter requires that the reader is familiar with complex numbers. Section 3.5 introduces Fourier series on complex form, which is a very convenient rewriting of the series in terms of complex exponential functions; we will stick to this format in the rest of the chapter. Section 3.7 exhibits the relationship between the regularity of a function (i.e., how many derivatives it has) with the decay of the Fourier coefficients (i.e., the question of how fast the coefficients in the Fourier series tend to zero). Section 3.8 deals with best N-term approximation, which is a method to obtain efficient approximations of infinite series using as few terms as possible. Finally, Section 3.9 gives a short introduction to the Fourier transform, which is used to obtain integral representations of aperiodic functions.

The material in Section 3.7 and Section 3.8 is the key to a deeper understanding of wavelets and why we need them. With the presentation chosen in this book, a large part of the wavelet chapters can be read without knowledge of these sections; but they are necessary for an understanding of Section 5.1 and Section 5.7.

3.1 Fourier series

We now introduce a type of series representation, which is well suited for analysis of periodic functions. Consider a function f defined on $\mathbb{R}$, and with period 2π; that is, we assume that

$$f(x + 2\pi) = f(x), \quad x \in \mathbb{R}.$$

We further assume that f belongs to the vector space

$$L^2(-\pi, \pi) := \left\{ f : \int_{-\pi}^{\pi} |f(x)|^2 dx < \infty \right\}. \tag{3.1}$$

To such a function we associate formally (which we write "$\sim$") the series

$$f \sim \frac{1}{2}a_0 + \sum_{n=1}^{\infty}(a_n \cos nx + b_n \sin nx), \tag{3.2}$$

where

$$a_n = \frac{1}{\pi}\int_{-\pi}^{\pi} f(x)\cos nx \, dx, \quad n = 0, 1, 2, \ldots$$

and

$$b_n = \frac{1}{\pi}\int_{-\pi}^{\pi} f(x)\sin nx \, dx, \quad n = 1, 2, \ldots \, .$$

The assumption (3.1) implies that the integrals defining a_n and b_n are well defined.

The series (3.2) is called the *Fourier series* associated to f, and a_n, b_n are called *Fourier coefficients*. As before, the Nth partial sum of the Fourier series is

$$S_N(x) = \frac{1}{2}a_0 + \sum_{n=1}^{N}(a_n \cos nx + b_n \sin nx). \tag{3.3}$$

The expression (3.2) only introduces Fourier series on the formal level: so far we have not discussed whether the series is convergent. Convergence issues are actually quite delicate for Fourier series; we come back to this issue in Section 3.2. However, already now it is clear why we restrict attention to 2π-periodic functions: all the trigonometric functions appearing in the Fourier series are 2π-periodic, so if we want any pointwise relationship between f and the Fourier series, this assumption is necessary.

Calculation of the Fourier coefficients can often be simplified using the rules stated below:

(R1) If f is 2π-periodic, then

$$\int_0^{2\pi} f(x)dx = \int_a^{a+2\pi} f(x)dx, \ \forall a \in \mathbb{R}.$$

(R2) If f is even, i.e., $f(x) = f(-x)$ for all x, then

$$\int_{-a}^{a} f(x)dx = 2\int_0^a f(x)dx, \ \forall a > 0.$$

(R3) If f is odd, i.e., $f(x) = -f(-x)$ for all x, then

$$\int_{-a}^{a} f(x)dx = 0, \ \forall a > 0.$$

If f is an even function, then $x \mapsto f(x)\cos nx$ is even and $x \mapsto f(x)\sin nx$ is odd; if f is odd, then $x \mapsto f(x)\cos nx$ is odd and $x \mapsto f(x)\sin nx$ is even. If we combine these observations with the rules above, we obtain the following result:

Theorem 3.1.1 *If f is an even function, then $b_n = 0$ for all n, and*

$$a_n = \frac{2}{\pi}\int_0^{\pi} f(x)\cos nx \ dx, \ n = 0, 1, 2, \dots .$$

If f is odd, then $a_n = 0$ for all n, and

$$b_n = \frac{2}{\pi}\int_0^{\pi} f(x)\sin nx \ dx, \ n = 1, 2, \dots .$$

In words, Theorem 3.1.1 says that the Fourier series for an even function consists of cosine-functions, while for an odd function it consists of sine-functions.

Let us find the Fourier series for a few 2π-periodic functions. Note that for such functions, it is sufficient to specify its function values on an interval of length 2π, e.g., the interval $[-\pi, \pi[$.

Example 3.1.2 Consider the *step function*

$$f(x) = \begin{cases} -1 \text{ if } x \in [-\pi, 0[, \\ 0 \text{ if } x = 0, \\ 1 \text{ if } x \in]0, \pi[, \end{cases} \tag{3.4}$$

extended to a 2π-periodic function. The function f is odd, so via Theorem 3.1.1 we see that $a_n = 0$ for all n and that

$$\begin{aligned} b_n &= \frac{2}{\pi}\int_0^\pi f(x)\sin nx \, dx \\ &= \frac{2}{\pi}\int_0^\pi \sin nx \, dx \\ &= \begin{cases} 0 \text{ if } n \text{ is even,} \\ \frac{4}{n\pi} \text{ if } n \text{ is odd.} \end{cases} \end{aligned}$$

Thus, the Fourier series is

$$\begin{aligned} f &\sim \frac{1}{2}a_0 + \sum_{n=1}^{\infty}(a_n \cos nx + b_n \sin nx) \\ &= \sum_{n \text{ odd}} \frac{4}{n\pi} \sin nx \\ &= \frac{4}{\pi}(\sin x + \frac{1}{3}\sin 3x + \frac{1}{5}\sin 5x + \cdots) \\ &= \frac{4}{\pi}\sum_{n=1}^{\infty}\frac{1}{2n-1}\sin(2n-1)x. \end{aligned} \tag{3.5}$$

Figures 3.1.3–3.1.4 show that the partial sums approximate f well in points where f is continuous. The figure also shows that there is a problem in the neighborhood of points where f is discontinuous: all the considered partial sums are "shooting over the correct value". This turns out to be a general phenomena for Fourier series; it is called *Gibb's phenomena,* and appears for arbitrary Fourier series of functions with a jump. One can show that the overshooting is about 9% of the size of the jump, regardless of the considered function.

It is interesting to notice that the Fourier series (3.5) actually converges pointwise for all $x \in \mathbb{R}$; this follows from Theorem 3.2.3 stated in the next section, but is nontrivial to prove directly based on the results we have presented in the tutorial on infinite series. In order to illustrate the difficulties, we mention that the series $\sum_{n=1}^{\infty} \frac{1}{2n-1}$ is divergent; this can be derived from Example 2.2.3. So the convergence of (3.5) is a consequence of the terms $\frac{1}{2n-1}$ being multiplied with $\sin(2n-1)x$. We also note that (3.5) does not converge absolutely: in fact,

$$\sum_{n=1}^{\infty}\left|\frac{1}{2n-1}\sin(2n-1)x\right|$$

is clearly divergent for $x = \pi/2$. The conclusion of these observations is that the convergence of (3.5) is a consequence of the sign-changes in the term $\sin(2n-1)x$.

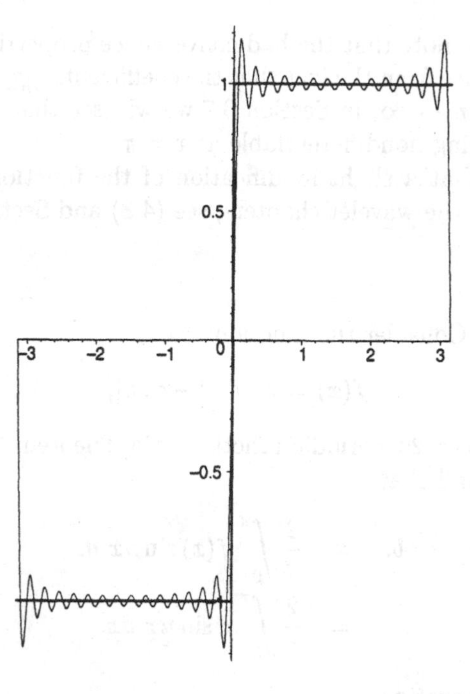

Figure 3.1.3 *The function f given by* (3.4) *and the partial sum S_{15}.*

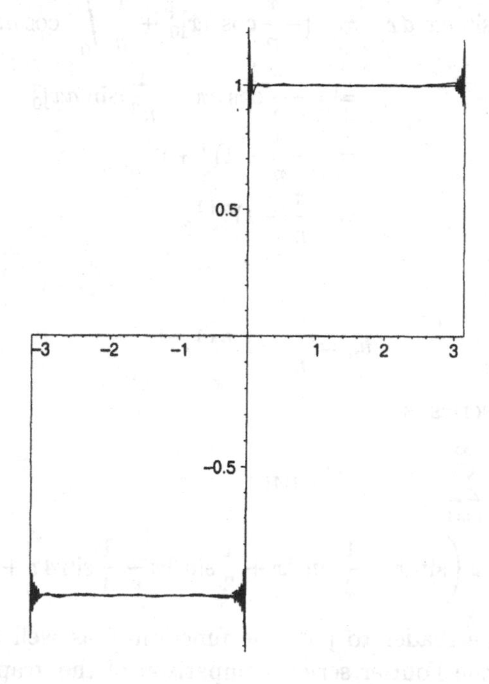

Figure 3.1.4 *The function f given by* (3.4) *and the partial sum S_{100}. Observe the overshooting of the partial sum around the points where f is discontinuous.*

For later use we note that the bad convergence properties of the Fourier series in (3.5) comes from the fact that the coefficients $\frac{1}{2n-1}$ only tend quite slowly to zero as $n \to \infty$. In Section 3.7 we will see that this is related to the function f being nondifferentiable at $x = \pi$.

We also note that a slight modification of the function f will play an important role in the wavelet chapters; see (4.8) and Section 5.4. □

Example 3.1.5 Consider the function

$$f(x) = x, \; x \in]-\pi, \pi[,$$

again extended to a 2π-periodic function. Via Theorem 3.1.1 we see that $a_n = 0$ for all n and that

$$\begin{aligned} b_n &= \frac{2}{\pi}\int_0^\pi f(x) \sin nx \, dx \\ &= \frac{2}{\pi}\int_0^\pi x \sin nx \, dx. \end{aligned}$$

Using partial integration,

$$\begin{aligned} \int_0^\pi x \sin nx \, dx &= [-\frac{x}{n}\cos nx]_0^\pi + \frac{1}{n}\int_0^\pi \cos nx \, dx \\ &= -\frac{\pi}{n}\cos n\pi + \frac{1}{n^2}[\sin nx]_0^\pi \\ &= -\frac{\pi}{n}(-1)^n + 0 \\ &= \frac{\pi}{n}(-1)^{n+1}. \end{aligned}$$

Thus,

$$b_n = \frac{2}{n}(-1)^{n+1},$$

and the Fourier series is

$$\begin{aligned} f &\sim \sum_{n=1}^{\infty} \frac{2}{n}(-1)^{n+1} \sin nx \qquad (3.6) \\ &= 2\left(\sin x - \frac{1}{2}\sin 2x + \frac{1}{3}\sin 3x - \frac{1}{4}\sin 4x + \cdots\right). \end{aligned}$$

We encourage the reader to plot the function f as well as the first few partial sums of the Fourier series: comparison of the graphs again shows that the partial sums approximate f well, except around points where f is nondifferentiable. Theorem 3.2.3 in the next section will give a formal explanation of this. □

3.2 Fourier's theorem and approximation

We now turn to a discussion of pointwise convergence of Fourier series. Having our experience with power series in mind, it is natural to ask for the Fourier series for a function f to converge pointwise toward $f(x)$ for each $x \in \mathbb{R}$. However, without extra knowledge about the function, this is too optimistic:

Example 3.2.1 Let $f \in L^2(-\pi, \pi)$, and define the function $g \in L^2(-\pi, \pi)$ by

$$g(x) = f(x) \text{ if } x \notin \mathbb{Z}, \; g(x) = f(x) + 1 \text{ if } x \in \mathbb{Z}.$$

Since an integral is invariant under a change of the value of the integrand in a few points, f and g have exactly the same Fourier coefficients, and therefore the same Fourier series. So at least for one of the functions, the Fourier series can not converge pointwise to the function for $x \in \mathbb{Z}$. □

Let us mention an even worse example:

Example 3.2.2 Consider the function

$$f(x) = \begin{cases} 1 \text{ if } x \in]-\pi, \pi[\cap \mathbb{Q}, \\ 0 \text{ if } \in [-\pi, \pi[\setminus \mathbb{Q}. \end{cases}$$

Readers with knowledge of the Lebesgue integral can prove that all the Fourier coefficients for this function are zero. Thus the Fourier series equals zero, and does not converge to $f(x)$ if $x \in \mathbb{Q}$. □

These examples show that certain conditions are necessary if we want any pointwise relationship between a function and its Fourier series. It turns out that conditions on the smoothness of f will be sufficient in order to obtain such relationships.

A function f on $\mathbb{R}$ is said to be *piecewise differentiable* if f is differentiable with a continuous derivative on each bounded interval — except maybe at a finite number of points $x_0, x_1, \ldots, x_n$; in a point x_j where f is non-differentiable we also require that the limits

$$\lim_{x \to x_j^+} f(x), \; \lim_{x \to x_j^-} f(x), \; \lim_{x \to x_j^+} f'(x), \text{ and } \lim_{x \to x_j^-} f'(x)$$

exist. For functions satisfying these conditions we have the following important result:

Theorem 3.2.3 *Assume that f is piecewise differentiable and 2π-periodic. Then the Fourier series converges pointwise for all $x \in \mathbb{R}$. For the sum function we have the following:*

(i) If f is continuous in x, then

$$\frac{1}{2}a_0 + \sum_{n=1}^{\infty}(a_n \cos nx + b_n \sin nx) = f(x); \tag{3.7}$$

(ii) If x_j is a point of discontinuity for f, then

$$\frac{1}{2}a_0 + \sum_{n=1}^{\infty}(a_n \cos nx_j + b_n \sin nx_j) = \frac{1}{2}\left(\lim_{x \to x_j^+} f(x) + \lim_{x \to x_j^-} f(x)\right).$$

Now assume that f is continuous everywhere. Then the Fourier series converges uniformly to f, and the maximal deviation between $f(x)$ and the partial sum $S_N(x)$ can be estimated by

$$|f(x) - S_N(x)| \leq \frac{1}{\sqrt{N}}\frac{1}{\sqrt{\pi}}\sqrt{\int_{-\pi}^{\pi}|f'(t)|^2 dt}. \tag{3.8}$$

Theorem 3.2.3 often allows us to get rid of the "mysterious sign $\sim$" in the definition of Fourier series: in words, it says that for reasonable functions, the sign "$\sim$" can be replaced by "=", except at points where the function is discontinuous.

The first part of Theorem 3.2.3, that is, the statement about the convergence of Fourier series, is called *Fourier's theorem.* Fourier published the result in 1822 in the article [13], actually with the claim that the Fourier series converges without any assumption on the function. He gave quite intuitive reasons for his claim, and the article does not reach the level of precision required for mathematical publications nowadays. Already around the time of its publication, many researchers protested against the result, and later it was proved that the result does not hold as generally as Fourier believed. In fact, things can go very bad if we do not impose the conditions in Theorem 3.2.3: there exist functions for which all the Fourier coefficients are well defined, but for which the Fourier series does not converge at any point! These functions, however, do not belong to $L^2(-\pi, \pi)$. For functions in $L^2(-\pi, \pi)$ the Fourier series converges pointwise almost everywhere; the exact meaning of "almost everywhere" can be found in textbooks on measure theory.

When this is said, we should also add that one has to admire Fourier for his intuition. Basically he was right, and his claim had a strong influence on development of mathematics: a large part of the mathematics from the nineteenth and twentieth century was invented in the process of finding the conditions for convergence of Fourier series and applying the results to, e.g., solution of differential equations.

The importance of Theorem 3.2.3 lies in the fact that it shows how a large class of functions can be decomposed into a sum of elementary sine and cosine functions. As a more curious consequence we mention that this frequently allows us to determine the exact sum of certain infinite series:

Example 3.2.4 For the step function in (3.4), Example 3.1.2 implies that

$$f(x) = \frac{4}{\pi} \sum_{n=1}^{\infty} \frac{1}{2n-1} \sin(2n-1)x, \; \forall x \in \mathbb{R}. \tag{3.9}$$

For $x \notin \pi\mathbb{Z} := \{0, \pm\pi, \pm 2\pi, \dots\}$, this result follows from Theorem 3.2.3 (i). For $x \in \pi\mathbb{Z}$ it follows from (ii) in the same theorem and the special definition of $f(0)$: a different choice of $f(0)$ would not change the Fourier series, but (3.9) would no longer hold for $x \in \pi\mathbb{Z}$.

Applying (3.9) with $x = \pi/2$ shows that

$$1 = \frac{4}{\pi} \sum_{n=1}^{\infty} \frac{1}{2n-1} \sin(2n-1)\frac{\pi}{2} = \frac{4}{\pi} \sum_{n=1}^{\infty} \frac{1}{2n-1}(-1)^{n-1},$$

or,

$$\sum_{n=1}^{\infty} \frac{1}{2n-1}(-1)^{n-1} = \frac{\pi}{4}. \qquad \square$$

For a continuous function f, the assumptions in Theorem 3.2.3 imply that the Fourier series converges uniformly to f. Equation (3.8) can be used to estimate how many terms we need to keep in the Fourier series in order to guarantee a certain approximation of the function f: if we want that $|f(x) - S_N(x)| \leq \epsilon$ for a certain $\epsilon > 0$, we can choose N such that

$$\frac{1}{\sqrt{N}} \frac{1}{\sqrt{\pi}} \sqrt{\int_{-\pi}^{\pi} |f'(t)|^2 dt} \leq \epsilon,$$

i.e.,

$$N \geq \frac{\int_{-\pi}^{\pi} |f'(t)|^2 dt}{\pi \epsilon^2}. \tag{3.10}$$

Note that (3.10) is a "worst case estimate": it gives a value of $N \in \mathbb{N}$ which can be used for all functions satisfying the conditions in Theorem 3.2.3. In order to minimize the calculation cost we usually want to obtain a given approximation using as small values of N as possible. In concrete cases where the Fourier coefficients are known explicitly, the next result can often be used to prove that smaller values of N than suggested in (3.10) are sufficient.

Proposition 3.2.5 *Assume that f is continuous, piecewise differentiable and 2π-periodic, with Fourier coefficients a_n, b_n. Then*

$$|f(x) - S_N(x)| \leq \sum_{n=N+1}^{\infty} (|a_n| + |b_n|), \ \forall x \in \mathbb{R}. \tag{3.11}$$

Proof: By Theorem 3.2.3, the assumptions imply that the Fourier series converges to $f(x)$ for all $x \in \mathbb{R}$. Via (3.7) and (3.3),

$$\begin{aligned}
|f(x) - S_N(x)| &= \left| \frac{1}{2}a_0 + \sum_{n=1}^{\infty}(a_n \cos nx + b_n \sin nx) \right. \\
&\quad \left. - \left(\frac{1}{2}a_0 + \sum_{n=1}^{N}(a_n \cos nx + b_n \sin nx) \right) \right| \\
&= \left| \sum_{n=N+1}^{\infty} (a_n \cos nx + b_n \sin nx) \right| \\
&\leq \sum_{n=N+1}^{\infty} |a_n \cos nx + b_n \sin nx| \\
&\leq \sum_{n=N+1}^{\infty} (|a_n| + |b_n|).
\end{aligned}$$

□

In the next example we compare Theorem 3.2.3 and Proposition 3.2.5.

Example 3.2.6 Consider the 2π-periodic function given by

$$f(x) = |x|, \ x \in [-\pi, \pi[.$$

Our purpose is to find estimates for $N \in \mathbb{N}$ such that

$$|f(x) - S_N(x)| \leq 0.1 \text{ for all } x \in \mathbb{R}. \tag{3.12}$$

The reader can check (Exercise 3.10) that the Fourier series of f is given by

$$f \sim \frac{\pi}{2} - \frac{4}{\pi} \sum_{n=1}^{\infty} \frac{1}{(2n-1)^2} \cos(2n-1)x.$$

See Figures 3.2.7–3.2.8, which show the function f and some partial sums. According to Theorem 3.2.3, the Fourier series converges uniformly to f.

We first apply (3.10), which was derived as a consequence of Theorem 3.2.3: it shows that (3.12) is satisfied if

$$N \geq \frac{2\pi}{0.1^2\pi} = 200.$$

Let us now apply Proposition 3.2.5. First, for any $k \in \mathbb{N}$, the partial sum

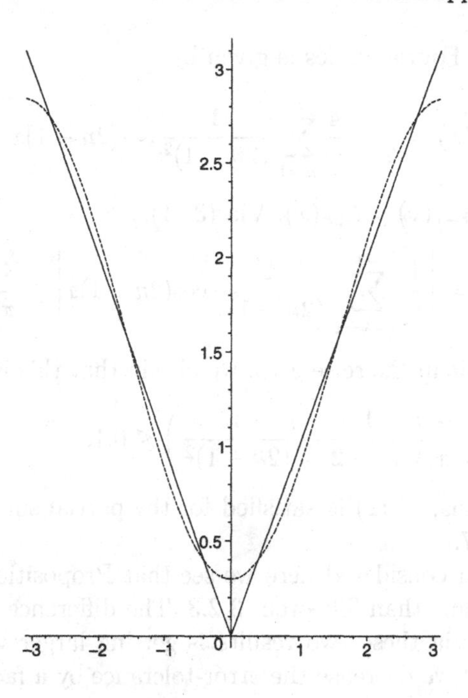

Figure 3.2.7 *The function $f(x)) = |x|$ and the partial sum $S_3(x)$, shown on the interval $[-\pi, \pi]$.*

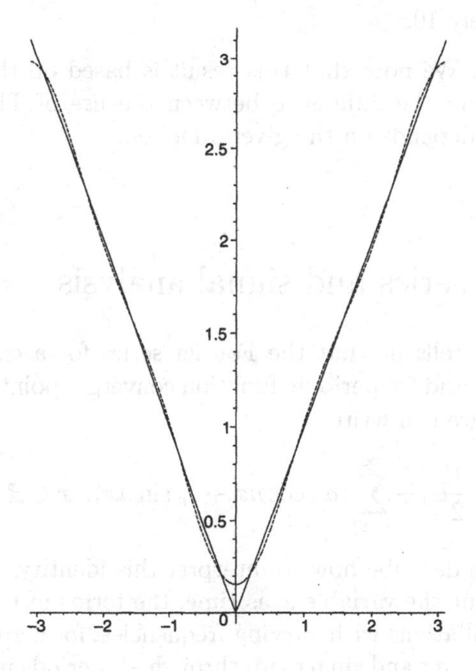

Figure 3.2.8 *The function $f(x)) = |x|$ and the partial sum $S_5(x)$, shown on the interval $[-\pi, \pi]$.*

$S_{2k-1}(x)$ of the Fourier series is given by

$$S_{2k-1}(x) = \frac{\pi}{2} - \frac{4}{\pi}\sum_{n=1}^{k}\frac{1}{(2n-1)^2}\cos(2n-1)x.$$

Note also that $S_{2k-1}(x) = S_{2k}(x)$. Via (3.11),

$$|f(x) - S_{2k-1}(x)| = \left|\frac{4}{\pi}\sum_{n=k+1}^{\infty}\frac{1}{(2n-1)^2}\cos(2n-1)x\right| \leq \frac{4}{\pi}\sum_{n=k+1}^{\infty}\frac{1}{(2n-1)^2}.$$

Applying the result in Exercise 2.13, we obtain that this is satisfied if

$$\frac{4}{\pi}\left(\frac{1}{4k+2} + \frac{1}{(2k+1)^2}\right) \leq 0.1,$$

i.e., for $k \geq 4$. Thus, (3.12) is satisfied for the partial sum $S_N(x)$ as soon as $N \geq 2k - 1 = 7$.

For the function considered here we see that Proposition 3.2.5 leads to a much better result than Theorem 3.2.3. The difference between the estimates obtained via these two results is getting larger when we ask for better precision: if we decrease the error-tolerance by a factor of 10,

- Theorem 3.2.3 will increase the value of N by a factor of 100;
- Proposition 3.2.5 will increase the value of N by a factor of approximately 10.

See Exercise 3.10. We note that this result is based on the choice of the considered function: the difference between the use of Theorem 3.2.3 or Proposition 3.2.5 depends on the given function. □

3.3 Fourier series and signal analysis

Fourier's theorem tells us that the Fourier series for a continuous piecewise differentiable and 2π-periodic function converges pointwise toward the function; that is, we can write

$$f(x) = \frac{1}{2}a_0 + \sum_{n=1}^{\infty}(a_n \cos nx + b_n \sin nx), \quad x \in \mathbb{R}. \tag{3.13}$$

Our aim is now to describe how to interpret this identity.

If we think about the variable x as time, the terms in the Fourier series correspond to oscillations with varying frequencies: for a given value of $n \in \mathbb{N}$, the functions $\cos nx$ and $\sin nx$ run through $\frac{n}{2\pi}$ periods in a time interval of unit length, i.e., they correspond to oscillations with frequency $\nu = \frac{n}{2\pi}$. Now, given a function f, the identity (3.13) represents f as a superposition of harmonic oscillations with various frequencies; the coefficients in (3.13) clarify which oscillations are present in f, and what the amplitudes are (i.e., the "magnitude" of the oscillations). Let us make this clearer by considering an example.

Example 3.3.1 A harmonic oscillation with frequency $\frac{100}{\pi}$ can be written in the form

$$f(x) = A\cos(200x + \varphi),$$

where A denotes the amplitude and φ is the phase (describing "when the oscillation starts"). We can think of this signal as the tone produced by a musical instrument. Using standard trigonometric formulas, we can rewrite the expression for f as

$$f(x) = A\cos\varphi\cos(200x) - A\sin\varphi\sin(200x);$$

this expression equals the Fourier series for f, i.e., we have

$$a_{200} = A\cos\phi,\ b_{200} = -A\sin\phi,\ \text{and}\ a_n = b_n = 0\ \text{for}\ n \neq 200.$$

The Fourier series tells us that only the frequency $\nu = \frac{100}{\pi}$ is present in the signal. More generally, if we consider a signal made up of several oscillations with varying amplitudes and phases, the Fourier series will again reveal exactly which frequencies are present; for example, if the signal consists of several tones played with different forces and with varying phases, the Fourier series will express this. Note, however, that we are speaking about quite abstract musical instruments: real instruments of course produce tones which begin and end at a certain time, i.e., the signals are not periodic considered as functions on $\mathbb{R}$. They furthermore produce higher harmonics, i.e., oscillations with frequencies which are integer-multiples of the tone itself. Real musical signals should rather be analyzed via the Fourier transform, as discussed in Section 3.9. □

Fourier analysis plays a central role in signal analysis. Very often, it is used to extract certain information from a given signal. As an imaginary (but realistic) example, we can think about a signal which measures the vibrations of a certain device such as, e.g., an engine. Due to mechanical limitations, it can be that we know in advance that appearance of certain frequencies with large amplitudes in the vibrations will make the engine break down; this is a case where it is important to be able to analyze the frequency content of the signal. For a much deeper and detailed exposition of signal analysis we refer to, e.g., the textbook [23].

3.4 Fourier series and Hilbert spaces

From our description so far, it is not clear that there is a link between Fourier series and (infinite-dimensional) linear algebra; the purpose of this section is to show that this is indeed the case. On the technical level, this section is not needed for the further development of the book, except for Sections 5.1–5.9; a reader without the required background might skip it.

Before we relate Fourier series and linear algebra, let us consider a finite-dimensional vector space V equipped with an inner product $\langle \cdot, \cdot \rangle$; in case the vector space is complex, we choose the inner product to be linear in the first entry. Recall that a family of vectors $\{e_k\}_{k=1}^n$ belonging to V is an *orthonormal basis* for V if

(i) $\text{span}\{e_k\}_{k=1}^n = V$, i.e., each $f \in V$ has a representation as a linear combination of the vectors e_k; and

(ii)

$$\langle e_k, e_j \rangle = \begin{cases} 1 \text{ if } k = j, \\ 0 \text{ if } k \neq j. \end{cases}$$

If $\{e_k\}_{k=1}^n$ is an orthonormal basis for V, then each element $f \in V$ has a representation

$$f = \sum_{k=1}^{n} \langle f, e_k \rangle e_k. \tag{3.14}$$

A large part of the theory for linear algebra has an extension to a class of infinite-dimensional vector spaces called *Hilbert spaces.* A Hilbert space $\mathcal{H}$ is defined exactly as a vector space with an inner product $\langle \cdot, \cdot \rangle$ (and associated norm $|| \cdot || = \sqrt{\langle \cdot, \cdot \rangle}$), with an extra condition added to ensure that "things go nicely" in the infinite-dimensional case.

Before we give the formal definition, we mention that the definition of convergence for a sequence of numbers has an analogue in any vector space $\mathcal{H}$ with an inner product, or, more generally, in any vector space equipped with a norm $|| \cdot ||$. In fact, we say that a sequence $\{x_k\}_{k=1}^{\infty} \subset \mathcal{H}$ converges to $x \in \mathcal{H}$ if

$$||x - x_k|| \to 0 \text{ as } k \to \infty.$$

Also, in analogy with sequences in $\mathbb{R}$ or $\mathbb{C}$, we say that a sequence $\{x_k\}_{k=1}^{\infty}$ belonging to a normed vector space $\mathcal{H}$ is a *Cauchy sequence* if for each given $\epsilon > 0$ we can find a number $N \in \mathbb{N}$ such that

$$||x_k - x_\ell|| \leq \epsilon \text{ whenever } k, \ell \geq N. \tag{3.15}$$

In the case of the normed vector spaces $\mathbb{R}$ or $\mathbb{C}$, we know that a sequence is convergent if and only if it is a Cauchy sequence; that is, we can use the criterion (3.15) to check whether a sequence is convergent or not, without involving a potential limit. In general infinite-dimensional normed vector spaces, the situation is different: there might exist Cauchy sequences which are not convergent. Hilbert spaces are introduced in order to avoid this type of complication:

Definition 3.4.1 *A Hilbert space is a vector space equipped with an inner product* $\langle \cdot, \cdot \rangle$, *with the property that each Cauchy sequence is convergent.*

Every finite-dimensional vector space with an inner product is a Hilbert space; and one can view Hilbert spaces as an extension of this framework, where we now allow certain infinite-dimensional vector spaces. In Hilbert spaces, several of the main results from linear algebra still hold; however, in many cases, extra care is needed because of the infinitely many dimensions. The exact meaning of this statement will be clear soon.

In Hilbert spaces, one can also define orthonormal bases, which now become infinite sequences $\{e_k\}_{k=1}^{\infty}$; one characterization is that $\{e_k\}_{k=1}^{\infty}$ is an orthonormal basis for $\mathcal{H}$ if

$$||f||^2 = \sum_{k=1}^{\infty} |\langle f, e_k \rangle|^2, \ \forall f \in \mathcal{H} \ \text{ and } \ \langle e_k, e_j \rangle = \begin{cases} 1 \text{ if } k = j, \\ 0 \text{ if } k \neq j. \end{cases} \tag{3.16}$$

If $\{e_k\}_{k=1}^{\infty}$ is an orthonormal basis for $\mathcal{H}$, then one can prove that each $f \in \mathcal{H}$ has the representation

$$f = \sum_{k=1}^{\infty} \langle f, e_k \rangle e_k. \tag{3.17}$$

At a first glance, this representation looks similar to (3.14), but some words of explanation are needed: in fact, the vectors $\{e_k\}_{k=1}^{\infty}$ belong to an abstract vector space, so we have not yet defined what an infinite sum like (3.17) should mean! The exact meaning appears by a slight modification of our definition of convergence for a series of numbers, namely that we measure the difference between f and the partial sums of the series in (3.17) in the norm $|| \cdot ||$ associated to our Hilbert space. That is, (3.17) means by definition that

$$\left\| f - \sum_{k=1}^{N} \langle f, e_k \rangle e_k \right\| \to 0 \text{ as } N \to \infty.$$

In our context, the important fact is that (if we identify functions which are equal almost everywhere) $L^2(-\pi, \pi)$ is a Hilbert space when equipped with the inner product

$$\langle f, g \rangle = \int_{-\pi}^{\pi} f(x)\overline{g(x)}dx, \ \ f, g \in L^2(-\pi, \pi). \tag{3.18}$$

Furthermore, one can check that the functions

$$\left\{\frac{1}{\sqrt{2\pi}}\right\} \bigcup \left\{\frac{1}{\sqrt{\pi}}\cos nx\right\}_{n=1}^{\infty} \bigcup \left\{\frac{1}{\sqrt{\pi}}\sin nx\right\}_{n=1}^{\infty} \tag{3.19}$$

form an orthonormal basis for $L^2(-\pi,\pi)$. Using (3.17) on this orthonormal basis leads to

$$f = \langle f, \frac{1}{\sqrt{2\pi}}\rangle \frac{1}{\sqrt{2\pi}} + \sum_{n=1}^{\infty} \langle f, \frac{1}{\sqrt{\pi}}\cos n\cdot\rangle \frac{1}{\sqrt{\pi}}\cos nx \tag{3.20}$$

$$+ \sum_{n=1}^{\infty} \langle f, \frac{1}{\sqrt{\pi}}\sin n\cdot\rangle \frac{1}{\sqrt{\pi}}\sin nx. \tag{3.21}$$

Calculation of the coefficients in this expansion leads to

$$f = \frac{1}{2}a_0 + \sum_{n=1}^{\infty}(a_n \cos nx + b_n \sin nx), \tag{3.22}$$

where a_n, b_n are the Fourier coefficients. But this is just the expression we used as definition of the Fourier series! This shows that the Fourier series for a function f actually is the expansion of f in terms of the orthonormal basis (3.19). This coincidence explains the troubles we have with pointwise convergence of Fourier series: viewed as a Hilbert space-identity, the exact meaning of the identity (3.22) is that

$$\left\| f - \frac{1}{2}a_0 - \sum_{n=1}^{N}(a_n \cos(n\cdot) + b_n \sin(n\cdot)) \right\| \to 0 \text{ as } N \to \infty,$$

i.e., that

$$\int_{-\pi}^{\pi} \left| f(x) - \frac{1}{2}a_0 - \sum_{n=1}^{N}(a_n \cos nx + b_n \sin nx)\right|^2 dx \to 0 \text{ as } N \to \infty; \tag{3.23}$$

this is different from requiring pointwise convergence of the Fourier series to $f(x)$.

The reader might wonder why Fourier series are introduced in such a way that the correct interpretation of convergence of Fourier series is the one stated in (3.23). We will explain this in two steps. First, since we are looking at functions f defined on an interval, it is not sufficient to have pointwise convergence: even if a series converges to $f(x)$ for all x in the interval, it might be that no partial sum approximates f well *over the entire interval;* see our discussion in Section 2.6. So we need a concept of convergence that takes all points in the interval into account simultaneously.

The next natural idea would be to require that

$$\int_{-\pi}^{\pi} \left| f(x) - \frac{1}{2}a_0 - \sum_{n=1}^{N}(a_n \cos nx + b_n \sin nx)\right| dx \to 0 \text{ as } N \to \infty; \tag{3.24}$$

this condition involves all x in the interval. Furthermore, geometrically it has the nice interpretation that the area of the set limited by the graphs of f, the partial sum S_N, and the line segments $x = \pm\pi$, tend to zero as $N \to \infty$. We note that the condition (3.24) is similar to (3.23); the difference is that (3.23) involves a *square* on the integrand. This square destroys the above geometric interpretation of the condition; but it has the advantage that Fourier series now are determined in terms of convergence in the space $L^2(-\pi, \pi)$, which is a Hilbert space. A reader with a deeper knowledge of Hilbert spaces will recognize the importance of this fact; one important feature is that the Hilbert space structure involves having an inner product; and this, in turn, guarantees that certain important minimization problems always have a unique solution; see, e.g., [27].

Furthermore, in a certain sense, the nice geometric interpretation of (3.24) is actually kept in our definition of convergence: one can prove that if (3.23) is satisfied, then (3.24) holds as well. See, e.g., Theorem 1.38 in [31].

As a final remark, we notice that (3.14) and (3.17) connect linear algebra with our description of signal transmission in Section 2.7: for example, (3.14) shows that the vector $f \in V$ is completely determined by the sequence of numbers $\{\langle f, e_k\rangle\}_{k=1}^n$. That is, one can transmit information about the vector f simply by sending this sequence.

3.5 Fourier series in complex form

Complex exponential functions give a convenient way to rewrite Fourier series. In fact, with our definition of Fourier series in Section 3.1 we have to keep track of the numbers a_n and b_n for $n \in \mathbb{N}$, as well as the special definition of a_0; as we will see, considering Fourier series in complex form will allow us to work with one set of coefficients $\{c_n\}_{n=-\infty}^{\infty}$, where all coefficients are given by a single formula.

Let us denote the imaginary unit number by i. In order to arrive at the announced expression for the Fourier series, we first recall that the complex exponential function is defined by

$$e^{i\theta} = \cos\theta + i\sin\theta, \ \theta \in \mathbb{R}.$$

Replacing θ by $-\theta$ in this formula, a few manipulations lead to expressions for the basic trigonometric functions via complex exponentials:

$$\cos\theta = \frac{e^{i\theta} + e^{-i\theta}}{2}, \quad \sin\theta = \frac{e^{i\theta} - e^{-i\theta}}{2i}. \tag{3.25}$$

Let us now insert these expressions in the formula for the nth term, $n \neq 0$, in the Fourier series for a function $f \in L^2(-\pi, \pi)$:

$$\begin{aligned}
a_n \cos nx + b_n \sin nx &= \frac{1}{\pi}\left(\int_{-\pi}^{\pi} f(x)\cos nx \, dx\right)\frac{e^{inx}+e^{-inx}}{2} \\
&\quad + \frac{1}{\pi}\left(\int_{-\pi}^{\pi} f(x)\sin nx \, dx\right)\frac{e^{inx}-e^{-inx}}{2i} \\
&= \left(\frac{1}{2\pi}\int_{-\pi}^{\pi} f(x)(\cos nx - i\sin nx)dx\right)e^{inx} \\
&\quad + \left(\frac{1}{2\pi}\int_{-\pi}^{\pi} f(x)(\cos nx + i\sin nx)dx\right)e^{-inx} \\
&= \left(\frac{1}{2\pi}\int_{-\pi}^{\pi} f(x)e^{-inx}dx\right)e^{inx} \\
&\quad + \left(\frac{1}{2\pi}\int_{-\pi}^{\pi} f(x)e^{inx}dx\right)e^{-inx}.
\end{aligned}$$

Also,

$$\frac{1}{2}a_0 = \frac{1}{2\pi}\int_{-\pi}^{\pi} f(x)dx.$$

Let us now define the numbers

$$c_n := \frac{1}{2\pi}\int_{-\pi}^{\pi} f(x)e^{-inx}dx, \quad n \in \mathbb{Z}; \tag{3.26}$$

with this definition, the Nth partial sum of the Fourier series can be written as

$$\begin{aligned}
S_N(x) &= \frac{1}{2}a_0 + \sum_{n=1}^{N}(a_n \cos nx + b_n \sin nx) \\
&= \sum_{n=-N}^{N} c_n e^{inx}.
\end{aligned}$$

Thus, the partial sums of the Fourier series can be expressed using complex exponential functions instead of cosine and sine functions.

When speaking about the *Fourier series of f in complex form*, we simply mean the infinite series appearing in

$$f \sim \sum_{n=-\infty}^{\infty} c_n e^{inx}, \tag{3.27}$$

with the coefficients c_n defined as in (3.26). Mathematically, nothing has changed going from Fourier series in the original form to the complex form: an infinite series is defined via the finite partial sums, and for Fourier series in complex form, the partial sums are exactly the same as before, they are just written in a different way.

Note also that there is an easy way to come from the Fourier coefficients with respect to sine functions and cosine functions to the coefficients for the Fourier series in complex form; in fact,

$$c_0 = \frac{1}{2}a_0, \text{ and for } n \in \mathbb{N}, \ c_n = \frac{a_n - ib_n}{2}, \ c_{-n} = \frac{a_n + ib_n}{2}. \tag{3.28}$$

On the other hand, if we know the Fourier coefficients in complex form, we can find the coefficients a_n, b_n via

$$a_0 = 2c_0, \ a_n = c_n + c_{-n}, \ b_n = i(c_n - c_{-n}). \tag{3.29}$$

Let us finally relate the complex exponential functions to the topic of Section 3.4:

Theorem 3.5.1 *The functions* $\{\frac{1}{\sqrt{2\pi}}e^{imx}\}_{m\in\mathbb{Z}}$ *form an orthonormal basis for* $L^2(-\pi, \pi)$.

3.6 Parseval's theorem

So far, we have used Fourier series as a tool for approximation of functions. We now mention another type of result, which shows that we can use Fourier series to determine certain infinite sums. The key to this type of result is *Parseval's theorem*:

Theorem 3.6.1 *Assume that the function* $f \in L^2(-\pi, \pi)$ *has the Fourier coefficients* $\{a_n\}_{n=0}^{\infty}, \{b_n\}_{n=1}^{\infty}$, *or, in complex form,* $\{c_n\}_{n=-\infty}^{\infty}$. *Then*

$$\frac{1}{2\pi}\int_{-\pi}^{\pi} |f(x)|^2 dx = \frac{1}{4}|a_0|^2 + \frac{1}{2}\sum_{n=1}^{\infty}(|a_n|^2 + |b_n|^2) = \sum_{n=-\infty}^{\infty} |c_n|^2.$$

As a consequence of Parseval's theorem, we note that if $f \in L^2(-\pi, \pi)$, then

$$\sum_{n=-\infty}^{\infty} |c_n|^2 < \infty. \tag{3.30}$$

One can prove that the opposite holds as well: if (3.30) holds, then

$$f(x) := \sum_{n=-\infty}^{\infty} c_n e^{inx}$$

defines a function in $L^2(-\pi, \pi)$. Thus, we have a one-to-one correspondence between functions in $L^2(-\pi, \pi)$ and sequences satisfying (3.30). We will not go into a discussion of all the important theoretical implications of this result (see, e.g., [27]), but restrict ourselves to a few computational consequences of Theorem 3.6.1:

Example 3.6.2
(i) In Example 3.1.2, we saw that the Fourier series for the function

$$f(x) = \begin{cases} -1 \text{ if } x \in [-\pi, 0[, \\ 0 \text{ if } x = 0, \\ 1 \text{ if } x \in]0, \pi[\end{cases}$$

is

$$f \sim \frac{4}{\pi}\sum_{n=1}^{\infty} \frac{1}{2n-1} \sin(2n-1)x.$$

Now,

$$\int_{-\pi}^{\pi} |f(x)|^2 dx = \int_{-\pi}^{\pi} dx = 2\pi,$$

and

$$\frac{1}{4}|a_0|^2 + \frac{1}{2}\sum_{n=1}^{\infty}(|a_n|^2 + |b_n|^2) = \frac{1}{2}\sum_{n \text{ odd}} (\frac{4}{n\pi})^2 = \frac{8}{\pi^2}\sum_{n=1}^{\infty} \frac{1}{(2n-1)^2}.$$

By Parseval's theorem this implies that

$$\frac{1}{2\pi}\, 2\pi = \frac{8}{\pi^2}\sum_{n=1}^{\infty} \frac{1}{(2n-1)^2},$$

i.e.,

$$\sum_{n=1}^{\infty} \frac{1}{(2n-1)^2} = 1 + \frac{1}{3^2} + \frac{1}{5^2} + \cdots = \frac{\pi^2}{8}.$$

(ii) By Example 3.1.5, the Fourier series for the function

$$f(x) = x, \; x \in [-\pi, \pi[$$

is given by

$$f \sim \sum_{n=1}^{\infty} \frac{2}{n}(-1)^{n+1} \sin nx.$$

Furthermore,

$$\int_{-\pi}^{\pi} |f(x)|^2 dx = \int_{-\pi}^{\pi} x^2 dx = \frac{1}{3}[x^3]_{-\pi}^{\pi} = \frac{2\pi^3}{3}.$$

Now, by Parseval's theorem,

$$\begin{aligned} \frac{1}{2\pi}\,\frac{2\pi^3}{3} &= \frac{1}{2}\sum_{n=1}^{\infty} \left|\frac{2}{n}(-1)^{n+1}\right|^2 \\ &= 2\sum_{n=1}^{\infty} \frac{1}{n^2}. \end{aligned}$$

Therefore

$$\sum_{n=1}^{\infty} \frac{1}{n^2} = 1 + \frac{1}{2^2} + \frac{1}{3^2} + \cdots = \frac{\pi^2}{6}. \tag{3.31}$$

This result is very surprising: think about how the number π is introduced as the fraction between the circumference and the diameter in a circle. The result in (3.31) has been known for a long time, but an *explanation* of *why* π appears in this context was only given in the second half of the twentieth century by the Austrian mathematician Hlawka. □

Let us for a moment concentrate on Fourier series in complex form. Given a function $f \in L^2(-\pi, \pi)$ with Fourier coefficients $\{c_n\}_{n=-\infty}^{\infty}$, an interesting consequence of Parseval's theorem is that

$$\sum_{n=-\infty}^{\infty} |c_n|^2 < \infty.$$

In particular, this implies that

$$c_n \to 0 \text{ as } n \to \pm\infty.$$

However, c_n might tend quite slowly to zero, and nothing guarantees that $\sum_{n=-\infty}^{\infty} c_n$ is convergent. For example, one can show that there is a function $f \in L^2(-\pi, \pi)$ for which $c_0 = 0$ and $c_n = 1/|n|^{2/3}$ for $n \neq 0$; in this case Example 2.2.3 shows that $\sum_{n=-\infty}^{\infty} c_n$ is divergent. Another example for which $\sum_{n=-\infty}^{\infty} c_n$ is divergent will appear in Example 3.7.1. In applications of Fourier series one will always have to work with certain partial sums (computers can not handle infinite sums); and here it is a complication if the coefficients c_n tend slowly to zero, because it forces the engineers to include many terms in the partial sums in order to obtain a reasonable approximation. We return to this subject in the next section.

3.7 Regularity and decay of the Fourier coefficients

Motivated by the end of the last section, we now analyze how fast the Fourier coefficients c_n tend to zero as $n \to \pm\infty$. As already mentioned, we wish the coefficients to converge fast to zero because this, at least on the intuitive level and for reasonable functions, implies that the partial sums $S_N(x)$ of the Fourier series of f will look and behave similar to f already for small values of $N \in \mathbb{N}$.

One way of measuring how fast c_n converges to zero is to ask if we can find constants $C > 0$ and $p \in \mathbb{N}$ such that

$$|c_n| \leq \frac{C}{|n|^p} \text{ for all } n \neq 0; \tag{3.32}$$

if this is satisfied, we say that c_n *decays polynomially.*

We note that from a practical point of view, we want (3.32) to hold with as small a value of $C > 0$ as possible, and as large a value of $p \in \mathbb{N}$ as possible.

It turns out that the decay rate of the Fourier coefficients is strongly related with the smoothness of the given function. Before we state this result formally in Theorem 3.7.2, we consider a small example which lends strong support to this fact.

Example 3.7.1 In this example we consider the Fourier coefficients for the following two functions:

(i) $f(x) = 1, \ x \in]-\pi, \pi[$;

(ii)

$$f(x) = \begin{cases} -1 \text{ if } x \in [-\pi, 0[, \\ 0 \text{ if } x = 0, \\ 1 \text{ if } x \in]0, \pi[. \end{cases} \tag{3.33}$$

Note that the function in (ii) is a seemingly very innocent modification of the function in (i): instead of having f constant on the interval $]-\pi, \pi[$, we now allow a jump in the middle of the interval, i.e., the function is piecewise constant. As we will see, this modification has a drastic influence on the Fourier coefficients.

First, the function in (i) has the Fourier coefficients

$$c_0 = 1, \text{ and for } n \neq 0, \ c_n = 0;$$

that is, just by working with one Fourier coefficient, we can recover the function completely. The decay condition (3.32) is satisfied for any $C > 0, p \in \mathbb{N}$.

For the function in (ii), we calculated the Fourier coefficients in real form in Example 3.1.2; via (3.28),

$$c_0 = 0, \text{ and for } n \in \mathbb{N}, \ c_n = \begin{cases} 0 \text{ if } n \text{ is even}, \\ \frac{-2i}{n\pi} \text{ if } n \text{ is odd}, \end{cases} \quad c_{-n} = \begin{cases} 0 \text{ if } n \text{ is even}, \\ \frac{2i}{n\pi} \text{ if } n \text{ is odd}. \end{cases}$$

Thus, the decay condition (3.32) is only satisfied if we take $p = 1$ and $C \geq 2/\pi$, i.e., the Fourier coefficients decay quite slowly. Note also that $\sum_{n=-\infty}^{\infty} |c_n|$ is divergent; however, $\sum_{n=-\infty}^{\infty} |c_n|^2$ is convergent, in accordance with Parseval's theorem.

The example (ii) illustrates once more that it is problematic to approximate discontinuous functions with Fourier series: in an attempt to resolve the discontinuity, the Fourier series introduces oscillations with frequencies which do not appear in the analyzed function, and with quite large coefficients. □

In the following analysis we will use the Fourier series in complex form,

$$f \sim \sum_{n\in\mathbb{Z}} c_n e^{inx}, \text{ where } c_n = \frac{1}{2\pi}\int_{-\pi}^{\pi} f(x)e^{-inx}\,dx.$$

Similar results can be stated for Fourier series in real form, and proved via (3.29). We will only consider continuous, piecewise differentiable and 2π-periodic functions, for which Fourier's theorem says that the Fourier series converges pointwise to the function; that is, we can actually write

$$f(x) = \sum_{n\in\mathbb{Z}} c_n e^{inx}, \; x \in [-\pi, \pi[.$$

Theorem 3.7.2 *Assume that f is 2π-periodic, continuous, and piecewise differentiable. Then the following holds:*

(i) Assume that f is p times differentiable for some $p \in \mathbb{N}$ and that $f^{(p)}$ is bounded, i.e., there is a constant $D > 0$ such that $|f^{(p)}(x)| \leq D$ for all x. Then there exists a constant $C > 0$ such that

$$|c_n| \leq \frac{C}{|n|^p} \text{ for all } n \in \mathbb{Z}\setminus\{0\}.$$

(ii) Assume that for some $p \in \mathbb{N}$ there exists a constant $C > 0$ such that

$$|c_n| \leq \frac{C}{|n|^{p+2}} \text{ for all } n \in \mathbb{Z}\setminus\{0\}. \tag{3.34}$$

Then f is p times differentiable, and

$$f^{(p)}(x) = \sum_{n\in\mathbb{Z}} c_n (in)^p e^{inx}.$$

We prove Theorem 3.7.2 in Section A.5. Theorem 3.7.2 indicates that there is a relationship between the smoothness of a function and its content of oscillations with high frequencies (measured by the size of the Fourier coefficients for large values of n): the smoother the function is, the faster these coefficients tend to zero, i.e., the smaller the content of high-frequency oscillations in f is.

3.8 Best N-term approximation

As mentioned before, infinite series is a mathematical tool, aiming at exact representation of certain functions. When working with series representations in practice, we are only able to deal with finite sums; that is, even if a function f has an exact representation via, e.g., a Fourier series, we will have to let the computers work with the finite partial sums S_N for a certain value of $N \in \mathbb{N}$.

For Fourier series, the inequality (3.10) shows how to choose N such that the partial sum S_N approximates f sufficiently well (under the conditions in Theorem 3.2.3). Unfortunately, this expression shows that N becomes very large if we need a good approximation, i.e., if ϵ is small. We also see that if we reduce ϵ by a factor of ten, then N might increase by a factor of a hundred! In practice, this type of calculations is performed using computers, but it is actually a problem that the number of needed terms increases so fast with the desired precision. Many calculations hereby get complicated to perform, simply because they take too much time. One way of obtaining an improvement is to replace the partial sum S_N by another finite sum which approximates f equally well, but which can be expressed using fewer coefficients. This is the idea behind the subject *nonlinear approximation*.

We will not give a general presentation of nonlinear approximation, but focus on a special type, the so-called *best N-term approximation*.

The basic idea is that, given a function f, not all terms in S_N are "equally important". To be more exact, to find S_N amounts to calculating the numbers $a_0, \ldots, a_N, b_1, \ldots, b_N$ (or $c_{-N}, \ldots, c_N$ in complex form); however, some of these numbers might be very small and might not contribute in an essential way to S_N, or, in other words, to the approximation of f.

If our capacity limits us to calculate and store only N numbers, it is much more reasonable to pick the N coefficients which represent f "as well as possible". This usually corresponds to picking the N (numerically) largest coefficients in the Fourier series.

Best N-term approximation is a delicate issue for Fourier series when we speak about obtaining approximations in the pointwise sense. The reason is that picking the largest coefficients first amounts to a *reordering* of the Fourier series; since Fourier series do not always converge unconditionally, this might have an influence on the convergence properties. However, if

$$\sum_{n=1}^{\infty}(|a_n| + |b_n|) < \infty, \text{ or, in complex form, } \sum_{n=-\infty}^{\infty} |c_n| < \infty, \tag{3.35}$$

then Proposition 2.1.8 shows that the Fourier series converges unconditionally (in the pointwise sense), and this complication does not appear. Note that Theorem 3.7.2(i) specifies some classes of functions for which (3.35) is satisfied, e.g., by taking $p \geq 2$.

Best N-term approximation often takes place in some norms rather than in the pointwise sense. That is, given a certain norm on $L^2(-\pi, \pi)$, we want to find an index set $J \subset \mathbb{Z}$ containing N elements, such that $x \mapsto f(x) - \sum_{n \in J} c_n e^{inx}$ is as small as possible, measured in that norm. If we use the norm $||\cdot||$ associated to the inner product $\langle \cdot, \cdot \rangle$ in (3.18), then for any choice $J \subset \mathbb{Z}$,

$$\left\| f - \sum_{n \in J} c_n e^{in\cdot} \right\| = \left\| \sum_{n \notin J} c_n e^{in\cdot} \right\| = \left(2\pi \sum_{n \notin J} |c_n|^2 \right)^{1/2}$$

$$= \left(||f||^2 - 2\pi \sum_{n \in J} |c_n|^2 \right)^{1/2} ;$$

this shows that the optimal approximation (measured in that particular norm) indeed is obtained when $\{c_n\}_{n \in J}$ consists of the N numerically largest coefficients.

If a central part of the given signal contains high frequencies with large amplitudes, then best N-term approximation is an important issue. Consider for example the signal

$$f(x) = \cos(10^7 x).$$

For this signal the Fourier coefficients are

$$a_{10^7} = 1, \quad a_n = b_n = 0 \text{ otherwise};$$

that is, a truncation of the Fourier series will treat the signal as zero if it does not include sufficiently many terms. On the other hand, a best N-term approximation will immediately realize the existence of the component with the large frequency, and recover f completely already at the first step.

Let us finally explain why the name nonlinear approximation is associated with the method. In order to do so, let us first consider the traditional approximation, simply via the Nth partial sum in the Fourier series. Assume that two functions f, g are given, and that the Nth partial sums in their Fourier series are

$$\frac{1}{2}a_0 + \sum_{n=1}^{N} (a_n \cos nx + b_n \sin nx),$$

respectively,

$$\frac{1}{2}\alpha_0 + \sum_{n=1}^{N} (\alpha_n \cos nx + \beta_n \sin nx).$$

Then, the Nth partial sum in the Fourier series for the function $f + g$ is

$$\frac{1}{2}(a_0 + \alpha_0) + \sum_{n=1}^{N} ((a_n + \alpha_n) \cos nx + (b_n + \beta_n) \sin nx),$$

i.e., we obtain the coefficients in the Fourier expansion simply by adding the coefficients for f and g. With best N-term approximation, the situation is different: only in rare cases, the coefficients leading to the best approximation of $f + g$ appear by adding the coefficients giving the best N-term approximation of f and g: that is, there is no linear relationship between the given function and the coefficients. Let us illustrate this with an example:

Example 3.8.1 Let

$$f(x) = \cos x + \frac{1}{5}\sin x, \quad g(x) = -\cos x + \frac{1}{10}\sin x.$$

Then the best approximation of f, respectively, g, using one term of the Fourier series is

$$\tilde{f}(x) = \cos x, \text{ respectively, } \tilde{g}(x) = -\cos x.$$

Thus,

$$\tilde{f}(x) + \tilde{g}(x) = 0.$$

On the other hand,

$$(f+g)(x) = \frac{1}{5}\sin x + \frac{1}{10}\sin x;$$

the best approximation of this function via a single trigonometric function is

$$\frac{1}{5}\sin x.$$

□

3.9 The Fourier transform

As we have seen, Fourier series is a useful tool for representation and approximation of periodic functions via trigonometric functions. For aperiodic functions we need to search for other methods. The classical tool is the Fourier transform, which we introduce here; more recently, wavelets have entered the scene, and we come back to this in Chapters 4–5.

Fourier series expand 2π-periodic functions in terms of the trigonometric functions $\cos nx, \sin nx$; all of these have period 2π. Trigonometric functions $\cos \lambda x, \sin \lambda x$, where λ is not an integer, do not have period 2π; this makes it very reasonable that they do not appear in the Fourier series for a 2π-periodic function. We have also seen that it is natural to think of a Fourier series as a decomposition of the given function f into harmonic oscillations with frequencies $\frac{n}{2\pi}, n = 0, 1, ...$; the size of the contributions at these frequencies are given by the Fourier coefficients.

If we want to expand aperiodic functions, the situation is different: all frequencies can appear in the signal, and the discrete sum over the special frequencies $\frac{n}{2\pi}$ must be replaced with an integral over all frequencies. We will consider functions in the vector space

$$L^1(\mathbb{R}) = \left\{ f : \mathbb{R} \to \mathbb{C} \mid \int_{-\infty}^{\infty} |f(x)|\, dx < \infty \right\}.$$

For functions $f \in L^1(\mathbb{R})$, the *Fourier transform* is defined by

$$\hat{f}(\gamma) = \int_{-\infty}^{\infty} f(x) e^{-2\pi i x \gamma} dx.$$

Frequently it is also convenient to use the notation

$$(\mathcal{F}f)(\gamma) = \hat{f}(\gamma);$$

this notation indicates that we can look at the Fourier transform as an *operator*, which maps the function f to the function $\hat{f}$.

The Fourier transform indeed contains information about the frequency content of the function f: as we will see in Example 3.9.2, the Fourier transform of a function oscillating like a cosine function with frequency ν over a bounded interval has a peak at $\gamma = \nu$, i.e., the Fourier transform tells us that the signal contains an oscillation with this frequency. We shall derive this result as a consequence of a chain of rules for calculations with the Fourier transform, which we present below.

Before we proceed to these results, we mention that the Fourier transform can be defined on several function spaces other than $L^1(\mathbb{R})$. In wavelet analysis, it is important that the Fourier transform makes sense as a bijective mapping from $L^2(\mathbb{R})$, defined in (4.2), onto itself; we will not go into these details. For $f \in L^1(\mathbb{R})$, one can prove that $\hat{f}$ is a continuous function which tends to zero as $\gamma \to \pm\infty$.

Below we state some of the important rules for calculations with the Fourier transform:

Theorem 3.9.1 *Given $f \in L^1(\mathbb{R})$, the following holds:*

(i) If f is even, then

$$\hat{f}(\gamma) = 2 \int_0^{\infty} f(x) \cos(2\pi x \gamma) dx.$$

(ii) If f is odd, then

$$\hat{f}(\gamma) = -2i \int_0^{\infty} f(x) \sin(2\pi x \gamma) dx.$$

(iii) For $a \in \mathbb{R}$, let $(T_a f)(x) = f(x-a)$. Then the Fourier transform of the function $T_a f$ is

$$(\mathcal{F}T_a f)(\gamma) = \hat{f}(\gamma) e^{-2\pi i a \gamma}.$$

(iv) Let $\theta \in \mathbb{R}$. Then the Fourier transform of the function $g(x) = f(x)e^{2\pi i \theta x}$ is

$$(\mathcal{F}g)(\gamma) = \hat{f}(\gamma - \theta).$$

(v) If f is differentiable and $f' \in L^1(\mathbb{R})$, then

$$(\mathcal{F}f')(\gamma) = 2\pi i \gamma \hat{f}(\gamma).$$

Theorem 3.9.1 is proved in almost all textbooks dealing with the Fourier transform, and also in several textbooks concerning signal analysis; see, e.g., [23]. Note in particular the rules (ii) and (iii); in words rather than symbols, they say that

- taking the Fourier transform of a translated version of f is done by multiplying $\hat{f}$ with a complex exponential function;
- taking the Fourier transform of a function f which is multiplied with a complex exponential function, corresponds to a translation of $\hat{f}$.

Let us show how we can use some of these rules to find the Fourier transform of a cosine function on an interval:

Example 3.9.2 Given constants $a, \omega > 0$, we want to calculate the Fourier transformation of the function

$$f(x) = \cos(\omega x)\chi_{[-\frac{a}{2},\frac{a}{2}]}(x). \tag{3.36}$$

This signal corresponds to an oscillation which starts at the time $x = -a/2$ and last till $x = a/2$. If x is measured in seconds, we have $\frac{\omega}{2\pi}$ oscillations per second, i.e., the frequency is $\nu = \frac{\omega}{2\pi}$. In order to find the Fourier transform, we first look at the function $\chi_{[-\frac{a}{2},\frac{a}{2}]}$ separately. Since this is an even function, Theorem 3.9.1(i) shows that for $\gamma \neq 0$,

$$\begin{aligned}
\mathcal{F}\chi_{[-\frac{a}{2},\frac{a}{2}]}(\gamma) &= 2\int_0^\infty \chi_{[-\frac{a}{2},\frac{a}{2}]}(x)\cos(2\pi x\gamma)dx \\
&= 2\int_0^{\frac{a}{2}} \cos(2\pi x\gamma)dx \\
&= \frac{2}{2\pi\gamma}\left[\sin(2\pi x\gamma)\right]_{x=0}^{x=\frac{a}{2}} \\
&= \frac{\sin \pi a\gamma}{\pi\gamma}.
\end{aligned}$$

Returning to the function f, we can use (3.25) to write

$$f(x) = \frac{1}{2}e^{i\omega x}\chi_{[-\frac{a}{2},\frac{a}{2}]}(x) + \frac{1}{2}e^{-i\omega x}\chi_{[-\frac{a}{2},\frac{a}{2}]}(x).$$

Via Theorem 3.9.1(iii) this shows that

$$\begin{aligned}
\hat{f}(\gamma) &= \frac{1}{2}\mathcal{F}\chi_{[-\frac{a}{2},\frac{a}{2}]}(\gamma - \frac{\omega}{2\pi}) + \frac{1}{2}\mathcal{F}\chi_{[-\frac{a}{2},\frac{a}{2}]}(\gamma + \frac{\omega}{2\pi}) \\
&= \frac{1}{2}\left(\frac{\sin \pi a(\gamma - \omega/2\pi)}{\pi(\gamma - \omega/2\pi)} + \frac{\sin \pi a(\gamma + \omega/2\pi)}{\pi(\gamma + \omega/2\pi)}\right).
\end{aligned}$$

Figures 3.9.3–3.9.4 show $\hat{f}$ for $\omega = 20\pi$ and different values of a in (3.36). A larger value of a corresponds to the oscillation $\cos(\omega x)$ being present in the signal over a larger time interval; we see that this increases the peak of $\hat{f}$ at the frequency $\gamma = \nu = 10$. □

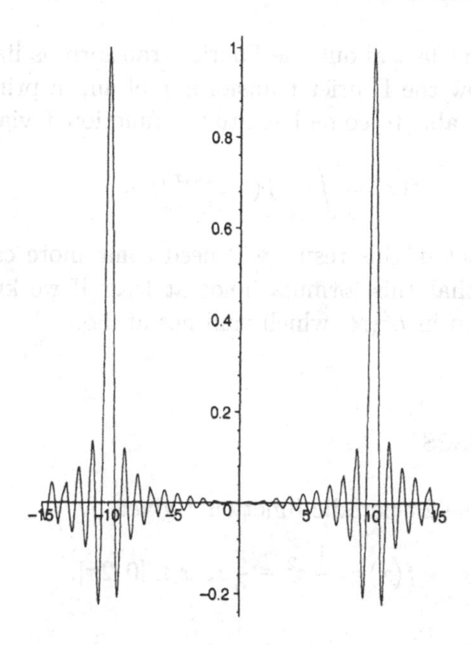

Figure 3.9.3 *The Fourier transform of the function f in (3.36) for $\omega = 20\pi, a = 2$; this corresponds to a signal with frequency $\nu = 10$ which is present during a time interval of length 2.*

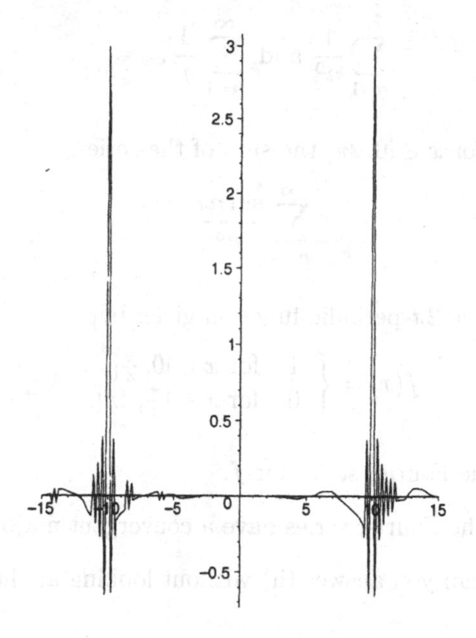

Figure 3.9.4 *The Fourier transform of the function f in (3.36) for $\omega = 20\pi, a = 6$; this corresponds to a signal with frequency $\nu = 10$ which is present during a time interval of length 6.*

A very important fact about the Fourier transform is its invertibility: if we happen to know the Fourier transform $\hat{f}$ of an, in principle, unknown function f, we are able to come back to the function f via

$$f(x) = \int_{-\infty}^{\infty} \hat{f}(\gamma)e^{2\pi ix\gamma}d\gamma. \tag{3.37}$$

An exact statement of this result will need some more care; for us, it is enough to know that this formula holds at least if we know that f is a continuous function in $L^1(\mathbb{R})$ which vanishes at $\pm\infty$.

3.10 Exercises

3.1 Consider the 2π-perodic function f given by

$$f(x) = \frac{1}{4}x^2 - \frac{\pi}{2}x, \; x \in [0, 2\pi].$$

(i) Prove that f is an even function.

(ii) Find the Fourier coefficients.

(iii) Find the sums of the series

$$\sum_{n=1}^{\infty} \frac{1}{n^2} \text{ and } \sum_{n=1}^{\infty} \frac{1}{n^4}.$$

(iv) Find for $x \in [0, 2\pi]$ the sum of the series

$$\sum_{n=1}^{\infty} \frac{\sin nx}{n^3}.$$

3.2 Let f be the 2π-periodic function given by

$$f(x) = \begin{cases} 1 & \text{for } x \in [0, \frac{\pi}{4}[, \\ 0 & \text{for } x \in [\frac{\pi}{4}, 2\pi[. \end{cases}$$

(i) Find the Fourier series for f.

(ii) Does the Fourier series have a convergent majorant series?

(iii) How can you answer (ii) without looking at the Fourier series?

3.3 Consider the 2π-periodic function f given by

$$f(x) = \begin{cases} \sin x & \text{if } \; 0 < x \leq \pi, \\ 0 & \text{if } \; \pi < x \leq 2\pi. \end{cases}$$

Denote the Nth partial sum of the Fourier series by S_N. How large do we need to choose N such that

$$|f(x) - S_N(x)| \leq 0.1, \ \forall x \in \mathbb{R}?$$

3.4 Consider the 2π-periodic function f given by

$$f(x) = e^{-x}, \ x \in [0, 2\pi[.$$

Find the partial sum S_2 of the Fourier series in real form (Hint: calculate the complex Fourier coefficients first!).

3.5 Consider the 2π-periodic function f given by

$$f(x) = \begin{cases} 0 & \text{for} \quad -\pi \leq x \leq 0, \\ 1 & \text{for} \quad 0 < x < \pi. \end{cases}$$

(i) Prove that the complex Fourier series of f is given by

$$f \sim \frac{1}{2} + \sum_{n \text{ odd}, n=-\infty}^{\infty} \frac{1}{\pi i n} e^{inx}.$$

(ii) Use (i) to calculate $\sum_{n=0}^{\infty} \frac{1}{(2n+1)^2}$.

3.6 Is $\sum_{n=1}^{\infty} (-1)^n \cos nx$ the Fourier series of a function in $L^2(-\pi, \pi)$?

3.7 Consider the 2π-periodic function f given by

$$f(x) = x, \ x \in [-\pi, \pi[.$$

Combining Example 3.1.5 and Theorem 3.2.3 we see that

$$x = \sum_{n=1}^{\infty} \frac{2}{n} (-1)^{n+1} \sin nx, \quad x \in]-\pi, \pi[. \tag{3.38}$$

(i) Find $\sum_{n=1}^{\infty} \frac{(-1)^{n+1}}{2n-1}$.

(ii) The function f is clearly differentiable for $x \in]-\pi, \pi[$. Are we allowed to differentiate (3.38) "under the sum sign," which would lead to

$$1 = 2 \sum_{n=1}^{\infty} (-1)^{n+1} \cos nx?$$

Compare with Theorem 2.6.6.

3.8 (i) Prove that

$$|\sin x| = \frac{2}{\pi} - \frac{4}{\pi}\sum_{n=1}^{\infty} \frac{\cos 2nx}{(2n-1)(2n+1)}, \quad \forall x \in \mathbb{R}.$$

(ii) Calculate the number $\sum_{n=1}^{\infty} \frac{1}{(2n-1)(2n+1)}$.

(iii) Calculate the number $\sum_{n=1}^{\infty} \frac{1}{(2n-1)^2(2n+1)^2}$.

(iv) Write the Fourier series for $|\sin(\cdot)|$ in complex form.

(v) Compare the decay of the coefficients in the Fourier series for $\sin(\cdot)$ and $|\sin(\cdot)|$; see Theorem 3.7.2.

(vi) Denote the Nth partial sum of the Fourier series for $|\sin(\cdot)|$ by S_N. Find N such that

$$|\,|\sin x| - S_N(x)| \leq 0.1, \ \forall x \in \mathbb{R}.$$

3.9 Consider the odd 2π-periodic function, which for $x \in [0, \pi]$ is given by

$$f(x) = \frac{\pi}{96}(x^4 - 2\pi x^3 + \pi^3 x).$$

(i) Find $f(-\frac{\pi}{2})$.

(ii) Prove that

$$f(x) = \sum_{n=1}^{\infty} \frac{\sin(2n-1)x}{(2n-1)^5}, \quad x \in \mathbb{R}$$

(hint: $\int_0^{\pi} (x^4 - 2\pi x^3 + \pi^3 x) \sin nx \, dx = 24\frac{1-(-1)^n}{n^5}$, $n \in \mathbb{N}$).

(iii) Prove that

$$|f(x) - \sin x| \leq 0.01, \ \forall x \in \mathbb{R}$$

(hint: use the integral test).

3.10 This exercise supplements Example 3.2.6.

(i) Prove that the Fourier series in Example 3.2.6 has the announced form, and that the sign "$\sim$" can be replaced by "$=$".

(ii) Argue for the expression for $S_{2k-1}(x)$ and for the fact that $S_{2k-1}(x) = S_{2k}(x)$.

(iii) Use the two methods in Example 3.2.6 to find estimates for $N \in \mathbb{N}$ such that $|f(x) - S_N(x)| \leq 0.01$ for all $x \in \mathbb{R}$.

4
Wavelets and Applications

In Section 2.5 we defined the sum function associated to a given family of functions $f_0, f_1, \dots$ defined on the same set. However, in practice sum functions frequently appear in a different way: a certain class of functions is given, and we want to find "simple functions" $f_0, f_1, \dots$ such that each function f in the class has an expansion

$$f(x) = \sum_{n=0}^{\infty} a_n f_n(x) \tag{4.1}$$

for some coefficients a_n. We note that this idea is similar to what we have seen in the context of power series and Fourier series: these cases correspond to the functions f_n being polynomials or trigonometric functions, respectively.

The motivation again comes from approximation theory: the functions f might be difficult to work with, but if (4.1) holds we might hope that the finite partial sums approximate f well, i.e., that for some $N \in \mathbb{N}$,

$$f(x) \approx \sum_{n=0}^{N} a_n f_n(x).$$

We also want that these finite partial sums are convenient to work with, which in practice implies that the functions $f_0, f_1, \dots$ should have a simple and convenient structure. These conditions are all satisfied for power series and Fourier series; however, these tools are only available for special classes of functions, namely, the analytic functions and the periodic functions, respectively. For functions in $L^2(\mathbb{R})$ we have the Fourier transform at our disposal; however, this is an integral representation, and not a representation via an infinite series.

In this chapter we introduce *wavelets*, a mathematical tool leading to representations of the type (4.1) for a large class of functions f. Wavelet theory is very new (about 20 years old), but has already proved useful in many contexts. In this chapter we are not aiming at a detailed treatment in the technical sense, but we will use the knowledge we have collected so far to give a description of some of the fundamental ideas, the history, and some applications.

We begin in Section 4.1 with a description of wavelet systems and their relationship to general infinite series. The section focusses on wavelets as a mathematical tool to obtain series expansions of functions in $L^2(\mathbb{R})$. Section 4.2 describes the role of wavelets in signal processing; and Section 4.3 follows this line by a concrete application, namely, the FBI's use of wavelets to store fingerprints electronically. More flexibility is gained using wavelet packets rather than wavelets; this is described in Section 4.4. Finally, Section 4.5 warns the reader that wavelets do not solve all types of problems, and mentions an alternative called Gabor systems.

Note that in Chapter 5 we return to several of the subjects introduced here and give a more technical description.

4.1 About wavelet systems

In the theory for Fourier series we have represented periodic functions via trigonometric functions. For aperiodic functions these representations are not available, and we have to look for other tools.

In signal analysis it is common to consider functions belonging to the vector space

$$L^2(\mathbb{R}) = \left\{ f : \mathbb{R} \to \mathbb{C} \mid \int_{-\infty}^{\infty} |f(x)|^2 dx < \infty \right\}. \tag{4.2}$$

Except for $f = 0$, functions in $L^2(\mathbb{R})$ are never periodic; in fact, for a periodic function the integral in (4.2) can not be finite. Nevertheless, it is possible to obtain series expansions like (4.1) of functions in $L^2(\mathbb{R})$, they are just of a type we have not met yet.

In order to describe the series expansions we aim at, i.e., how to choose the functions f_k in (4.1), we consider a function ψ, which could look like the function in Figure 4.1.1. To ψ we associate a family of functions $\psi_{j,k}, j, k \in \mathbb{Z}$, defined by

$$\psi_{j,k}(x) = 2^{j/2}\psi(2^j x - k), \; x \in \mathbb{R}.$$

In order to understand the relationship between these functions and the given function ψ we first consider $j = 0$. We observe that for $k = 0$,

$$\psi_{0,0}(x) = 2^0\psi(2^0x - 0) = \psi(x);$$

thus, the function $\psi_{0,0}$ is ψ itself. Furthermore, for $k \in \mathbb{Z}$,

$$\psi_{0,k}(x) = 2^0\psi(2^0x - k) = \psi(x - k).$$

Thus, the graph of the function $\psi_{0,k}$ appears by translating the graph of ψ by k units to the right; see Figure 4.1.2. In order to understand the role of the parameter j, we now put $k = 0$. Then

$$\psi_{j,0}(x) = 2^{j/2}\psi(2^jx).$$

As Figure 4.1.3–4.1.6 demonstrate, the functions $\psi_{j,0}$ are scaled versions of ψ: when j is a large positive number, the graph of ψ_j is a compressed version of the graph of ψ, while negative values of j lead to less localized versions of ψ.

Putting everything together, we see that the functions $\psi_{j,k}$ are scaled and translated versions of ψ. We say that the functions $\psi_{j,k}, j, k \in \mathbb{Z}$, form the *wavelet system* associated to the function ψ.

Our goal is to expand functions in $L^2(\mathbb{R})$ in terms of functions of the type $\psi_{j,k}$; that is, we want to obtain expansions like (4.1) with f_k replaced by $\psi_{j,k}$. Since we have two parameters in $\psi_{j,k}$, the expansion will be in terms of a *double sum.* It is intuitively clear that this can only be possible under certain conditions on the function ψ. So our first problem is to determine a function ψ such that every function f in $L^2(\mathbb{R})$ has a representation of the form

$$f(x) = \sum_{j\in\mathbb{Z}}\sum_{k\in\mathbb{Z}} c_{j,k}\psi_{j,k}(x) \tag{4.3}$$

for a certain choice of the coefficients $\{c_{j,k}\}$. The coefficients will of course depend on the given function f. At the moment, we will put the additional constraint on ψ that

$$\int_{-\infty}^{\infty} \psi_{j,k}(x)\overline{\psi_{j',k'}(x)}dx = \begin{cases} 1 \text{ if } j = j', k = k', \\ 0 \text{ otherwise.} \end{cases} \tag{4.4}$$

One can prove that this extra condition implies that if (4.3) is possible for all $f \in L^2(\mathbb{R})$, then the coefficients appearing in the expansion of an arbitrary f are unique, and we have a convenient expression for them: in fact,

$$c_{j,k} = \int_{-\infty}^{\infty} f(x)\overline{\psi_{j,k}(x)}dx. \tag{4.5}$$

The condition (4.4) puts extra restrictions on the choice of ψ: in the technical terms used in Section 3.4 it forces $\{\psi_{j,k}\}_{j,k\in\mathbb{Z}}$ to be an orthonormal basis for $L^2(\mathbb{R})$. In Section 5.8 we will see that the representation in (4.3)

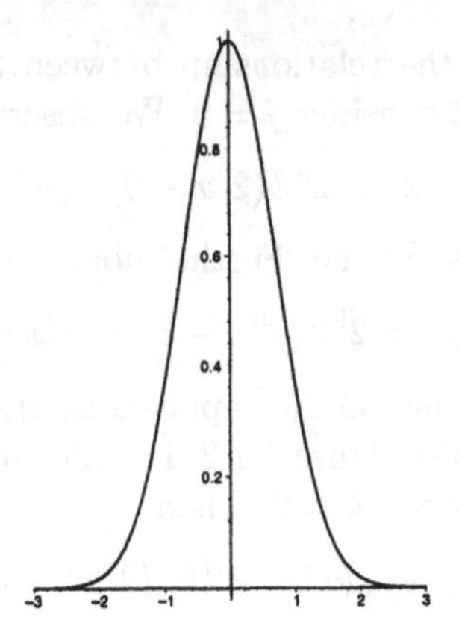

Figure 4.1.1 *The function* $\psi(x) = e^{-x^2} = \psi_{0,0}(x)$.

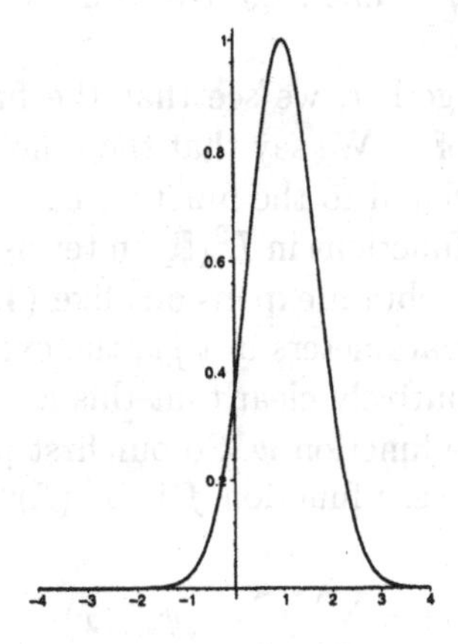

Figure 4.1.2 *The function* $\psi_{0,1}(x) = e^{-(x-1)^2}$.

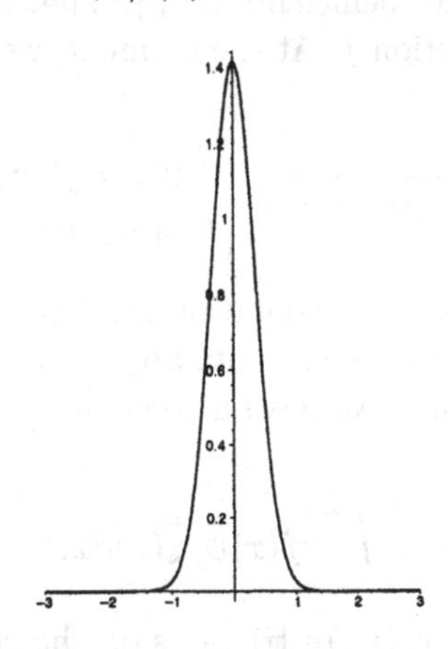

Figure 4.1.3 *The function* $\psi_{1,0}(x) = 2^{1/2}\psi(2x - 0) = 2^{1/2}e^{-4x^2}$.

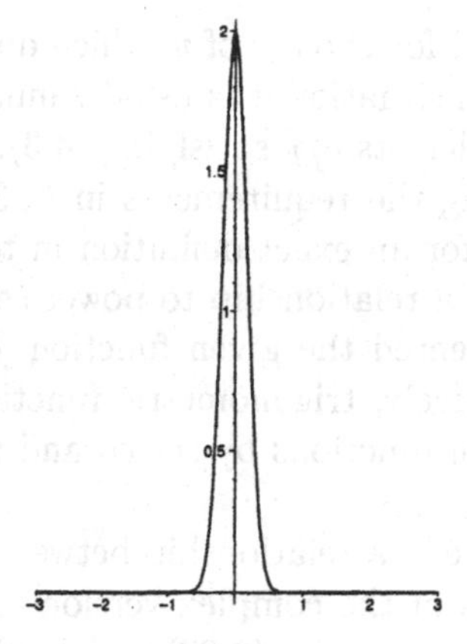

Figure 4.1.4 *The function* $\psi_{2,0}(x) = 2\psi(2^2x) = 2e^{-16x^2}$.

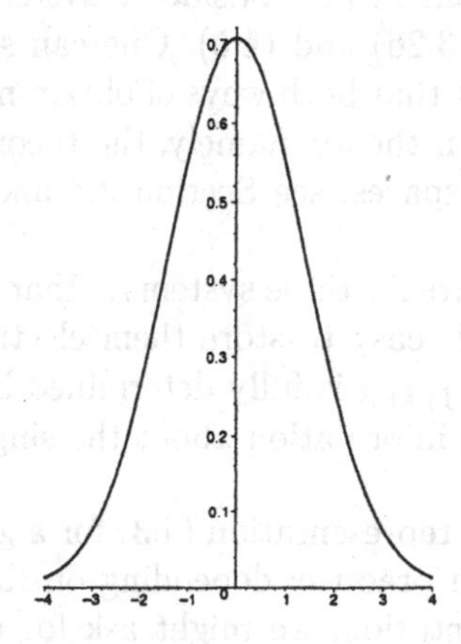

Figure 4.1.5 *The function* $\psi_{-1,0}(x) = 2^{-1/2}\psi(2^{-1}x) = \frac{1}{\sqrt{2}}e^{-x^2/4}$.

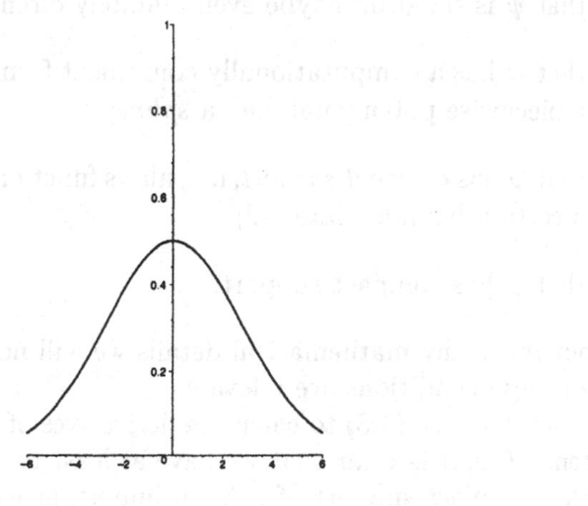

Figure 4.1.6 *The function* $\psi_{-2,0}(x) = 2^{-1}\psi(2^{-2}x) = \frac{1}{2}e^{-x^2/16}$.

very well can be obtained for choices of ψ which do not satisfy (4.4); however, in this more general situation it is usually much more complicated to find expressions for coefficients $c_{j,k}$ satisfying (4.3).

A function ψ satisfying the requirements in (4.3) and (4.4) is called a *wavelet*; see Section 5.1 for an exact definition in terms of Hilbert spaces. We immediately observe a relationship to power series and Fourier series: in these cases we represented the given function f as a superposition of power functions, respectively, trigonometric functions, and now we want to replace these classes of functions by scaled and translated versions of a fixed function.

Observe also that there is a relationship between the coefficients $c_{j,k}$ in (4.5) and the coefficients in the complex version of Fourier series: the coefficients in both of the expansions (3.27) and (4.3) appear by integration of the product of the function f we wish to represent, with the complex conjugates of the functions in the considered system of functions. See the concrete expressions in (3.26) and (4.5). One can show that these coincidences are due to the fact that both ways of obtaining series expansions are special cases of a common theory, namely, the theory for orthogonal function systems in Hilbert spaces; see Section 3.4 and Section 5.1 for more details on this subject.

Another common feature for these systems is that they have a convenient structure, which makes it easy to store them electronically. For example, the wavelet system $\{\psi_{j,k}\}_{j,k\in\mathbb{Z}}$ is fully determined by the function ψ: that is, we only need to store information about the single function ψ in order to be able to use (4.3).

The availability of the representation (4.3) for a given function f is usually not enough for us in practice: depending on the context in which we want to use the representation, we might ask for choices of ψ satisfying some extra conditions. Some of the conditions which appear often are

- that ψ is smooth, maybe even infinitely often differentiable;
- that ψ has a computationally convenient form; for example that ψ is a piecewise polynomial, i.e., a *spline*;
- that ψ has *compact support*, i.e., all its function values are zero outside a certain bounded interval;
- that $\hat{\psi}$ has compact support.

Without too many mathematical details we will now describe cases where some of these conditions are relevant.

If we want to use (4.3) to calculate derivatives of the function f, e.g., via Theorem 2.6.6, it is clear that we have at least to assume that ψ is differentiable. Compact support of ψ is an important issue for computer-based calculations; in fact, since a computer can handle only a finite collection of numbers, functions without compact support will always have to be truncated at some place.

In practice we can not expect that ψ has all the above properties, so we need to be careful and only insist on the properties that are needed in the application we have in mind. For example, it is impossible that both ψ and its Fourier transform have compact support; this is inconvenient in cases where we need to work with the time-behavior of functions as well as their frequency-content. In such cases, we must replace the wish for compact support of either ψ or $\hat{\psi}$ with the requirement that the function at least tends very fast to zero. Formulated for the function ψ, such a requirement could be that there exist constants $C, \alpha > 0$ such that

$$|\psi(x)| \leq Ce^{-\alpha|x|}, \ \forall x \in \mathbb{R}. \tag{4.6}$$

If ψ satisfies (4.6), we say that ψ decays *exponentially*.

As already mentioned, computer-based calculations in the context of series expansions always have to be based on finite partial sums; that is, the exact representation (4.3) has to be replaced by

$$f(x) \approx \sum_{j=-N}^{N} \sum_{k=-N}^{N} c_{j,k}\psi_{j,k}(x) \tag{4.7}$$

for a sufficiently large value of $N \in \mathbb{N}$. One of the advantages of wavelets (compared with other tools leading to signal expansions) is that often it is possible to obtain a good approximation of f using only a few coefficients. Not necessarily based on a small value of N in (4.7): maybe with a relatively large value of N but with most of the coefficients $\{c_{j,k}\}_{|j|,|k|\leq N}$ vanishing. As we have seen in Example 2.7.1, it is crucial to find partial sums which approximate the given signal well, and in the context of wavelets it facilitates the use of the approximation if only a few functions $\psi_{j,k}$ appear.

The first example of a function f satisfying (4.3) was presented by Haar in his Ph.D. thesis and published in the paper [16] from 1910:

Example 4.1.7 The *Haar wavelet* is the function given by

$$\psi(x) = \begin{cases} 1, & x \in [0, \frac{1}{2}[, \\ -1, & x \in [\frac{1}{2}, 1[, \\ 0, & \text{otherwise.} \end{cases} \tag{4.8}$$

The Haar wavelet satisfies the requirement that each $f \in L^2(\mathbb{R})$ has an expansion as in (4.3); the condition (4.4) is satisfied as well. Comparing with the list of desirable features on page 88, we note that ψ has compact support and is a piecewise polynomial. However, ψ is not continuous for $x = 0$, $x = \frac{1}{2}$, and $x = 1$. This might be a problem in the context of signal transmission. Recall from Section 2.7 that if we want to apply the representation (4.3) for signal transmission, the sender S will not be able to send the infinitely many numbers $c_{j,k}$, but only a finite set of coefficients, e.g., $\{c_{j,k}\}_{j,k=-N}^{N}$ for a certain value of N. Furthermore, roundoff errors might appear, so the receiver $\mathcal{R}$ will obtain a signal of the form

$$f \approx \sum_{j=-N}^{N} \sum_{k=-N}^{N} \tilde{c}_{j,k} \psi_{j,k}. \tag{4.9}$$

If ψ is the Haar wavelet, then even if f is an infinitely often differentiable function, $\mathcal{R}$ will receive a function which is not continuous. It is clear that this will have a negative effect on transmission of pictures with smooth boundaries: for the receiver, a picture of an elephant will appear to be composed of small boxes.

A direct calculation reveals in addition that the Fourier transform of ψ has quite slow decay. □

It took more than 70 years before Strömberg found the next wavelet, or rather, class of wavelets, in 1982:

Example 4.1.8 For each $k \in \mathbb{N}$, Strömberg constructed a wavelet which has derivatives up to order k; all the wavelets are splines, and they have exponential decay. □

From around 1985 there has been an intense research activity aiming at construction of wavelets with prescribed properties. As a rather curious example of how research happens, we mention that the search was initiated by the French mathematician Yves Meyer, who was sure that no wavelet could be infinitely often differentiable and decay exponentially. However, in the process of trying to prove his claim, he ended up realizing that he was wrong: in fact, he finally found a wavelet with the properties he thought were impossible. Today, this wavelet is called *Meyer's wavelet.*

The first wavelet constructions were considered small miracles: they were all based on some clever observations and tricks. For this reason it was believed that wavelets are so special that there would be very limited ways of finding them. In the late 1980s, this turned out to be wrong: Stéphane Mallat, a young engineer, and Yves Meyer found around 1987 a collection of conditions, which together lead to construction of wavelets. The result, namely, *multiresolution analysis,* turned very quickly into a standard tool in engineering and signal processing; see Section 5.2 for the definition and a few basic properties. The theory took a large step forward around 1990, where Ingrid Daubechies found ways to construct wavelets with compact support via multiresolution analysis: for $N \in \mathbb{N}$, the *Nth Daubechies' wavelet* equals zero outside an interval of the type $[0, 2N-1]$ (see Figure 4.1.9).

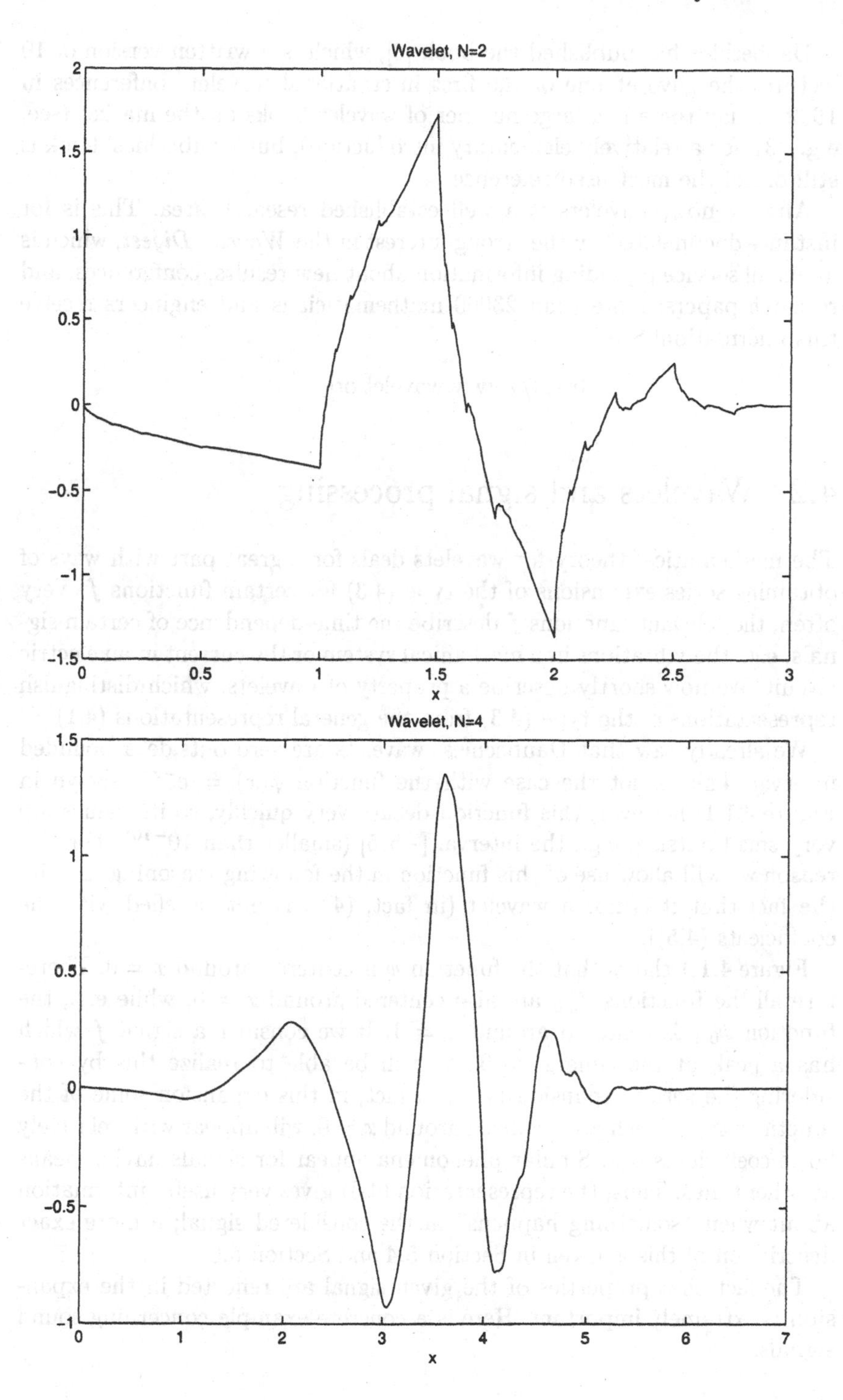

Figure 4.1.9 *Some of Daubechies' wavelets.*

Daubechies has published the book [8], which is a written version of 10 lectures she gave at one of the first international wavelet conferences in 1992. Today there is a large number of wavelet books on the market (see, e.g., [31] for a relatively elementary introduction), but Daubechies' book is still one of the most used references.

Already now, wavelets is a well established research area. This is for instance documented by the strong interest in the *Wavelet Digest,* which is an email service providing information about new results, conferences, and research papers: more than 23000 mathematicians and engineers receive this information! See

http://www.wavelet.org

4.2 Wavelets and signal processing

The mathematical theory for wavelets deals for a great part with ways of obtaining series expansions of the type (4.3) for certain functions f. Very often, the relevant functions f describe the time-dependence of certain signals, e.g., the vibrations in a mechanical system or the current in an electric circuit. We now shortly describe a property of wavelets, which distinguish representations of the type (4.3) from the general representations (4.1).

We already saw that Daubechies' wavelets are zero outside a bounded interval. This is not the case with the function $\phi(x) = e^{-x^2}$, shown in Figure 4.1.1; however, this function decays very quickly, so its values are very small outside, e.g., the interval $[-5, 5]$ (smaller than 10^{-10}). For this reason we will allow use of this function in the following reasoning, despite the fact that it is not a wavelet (in fact, (4.3) is not satisfied with the coefficients (4.5)).

Figure 4.1.1 shows that the function ψ is centered around $x = 0$. Therefore all the functions $\psi_{j,0}$ are also centered around $x = 0$, while e.g., the function $\psi_{0,1}$ is centered around $x = 1$. If we consider a signal f which has a peak at the time $x = 0$, we will be able to realize this by considering the series expansion (4.3): in fact, in this expansion some of the functions $\psi_{j,0}$, which are localized around $x = 0$, will appear with relatively large coefficients $c_{j,0}$. Similar phenomena appear for signals having peaks at other times. Thus, the representation (4.3) gives very useful information about when "something happens" in the considered signal; a more exact description of this is given in Section 5.4 and Section 5.6.

The fact that properties of the given signal are reflected in the expansion is extremely important. Here is a concrete example concerning sound signals:

Example 4.2.1 Wavelets are frequently used to remove noise from music recordings. The main idea is to think about a music signal as consisting of the music itself to which some noise is added. The music signal itself describes how the music changes in time; we can think about the signal as the current through the loudspeaker when we play a recording. Expanding the music piece via wavelets means that we represent this signal via the coefficients $\{c_{j,k}\}$ in (4.3); the coefficients "tell when something happens in the music". The noise contribution is usually small compared to the music, but irritating for the ears; it also contributes to the coefficients $\{c_{j,k}\}$, but usually less than the music itself. The idea is now to remove the coefficients in (4.3) which are smaller than a certain threshold value (strictly speaking, this procedure is not applied on the signal itself, but on its so called wavelet transform; see Section 5.4 for a more exact description). To be more precise, this means that these coefficients are replaced by zeros; thus, one lets the remaining coefficients represent the music. The idea is to remove the part of the signal which has no relationship to the music; unfortunately, the above procedure also might cancel smaller parts of the music itself, but still the result is usually considered by the ears as an improvement. □

As a curious fact we mention that this method has been applied on a recording dating back to 1889, where the composer and pianist Johannes Brahms plays one of his own piano pieces. Unfortunately, the recording is so noisy that it is even difficult to hear that it has anything to do with a piano. Using wavelet methods, J. Berger and C. Nichols managed to get rid of so much noise that it became possible to hear the music and obtain an idea about how Brahms played his own compositions. Somewhat surprisingly it turned out that he did not follow his own score very closely: he allowed himself several freedoms in the interpretation, for which a piano student would be corrected! See the original article [1] for a more detailed description.

In applications of (4.3), the relevant signals f are usually not given by concrete expressions in terms of elementary functions like x^n or e^x; they are rather the output from some electric device such as, e.g., a seismograph. We can split the class of signals we are interested in into two classes, namely:

- Signals described by a continuous function; for example a recording of a speech signal or music signal, which measures the current in the cable to the loudspeaker as a function of time; they are called *continuous signals.*
- Signals described by a sequence of numbers or pairs of numbers; they are called *discrete signals.*

Let us mention an example of a discrete signal:

Example 4.2.2 A digital black-white photo consists of a splitting of the picture into a large number of small squares, called *pixels;* to each pixel, the camera associates a light intensity, measured on a scale from, say, 0 (completely white) to 256 (completely black). Put together, this information constitutes the picture. Thus, mathematically a photo consists of a sequence of pairs of numbers, namely, a numbering of the pixels together with the associated light intensity. □

The wavelet theory presented so far is designed for analysis of continuous signals: for discrete signals it can not be applied directly. On the other hand, discrete signals are more suitable for processing in a computer. In fact, computers can not deal with functions like mathematicians do: they can only work with numbers, and as discussed before, basically all operations have to be reduced to the four elementary types of operations (addition, subtraction, multiplication, division). For this reason, computer based methods will always require some adaption of the theoretical results discussed so far.

Furthermore, there is a fundamental difference between the functions appearing in mathematics and in signal processing. Let us as an example consider the output from a seismograph, which measures the movements of the ground at a certain place. Due to, e.g., uncertainties of the measurements, it does not make much sense to claim that a given measurement from the seismograph took place *exactly* at the time 12:05:03; it is more reasonable to consider the data as an average over a certain small time interval. Thus, the output from the seismograph can be described by a sequence of numbers, which measure the average movement over consecutive small time intervals. Instead of thinking about these numbers as function values for a function on $\mathbb{R}$ and performing a wavelet analysis on $L^2(\mathbb{R})$, it is more convenient to let the processing of the signal work directly with the given numbers. Thus, one works directly with the finitely many pairs of numbers, where the first coordinate represents a small time interval, and the second coordinate the corresponding measurement.

Adapted to such finite sequences of numbers, certain discrete versions of wavelet theory have been developed. For example a number of methods for data compression have been investigated. Formulated rather sloppily, they perform some operations on the given numbers, which reduce the bits one has to store, while keeping the essential information in the signal. Let us give an example illustrating this. The example is taken from the book [22], which focusses on the type of wavelet analysis which is used in signal processing, namely, the ***discrete wavelet transformation.***

Example 4.2.3 Consider the following sequence of numbers (which represent a given signal)

56	40	8	24	48	48	40	16

To these eight numbers we now associate eight new numbers, which appear in the following way. First, we consider the above eight numbers as a series of four pairs of numbers, each containing two numbers. We now replace each of these pairs of numbers by two new numbers, namely, their *average* and the *difference* between the first number in the pair and the calculated average. The first pair of numbers in the given signal consists of the numbers 56 and 40; our procedure replaces them by the new pair consisting of the numbers $\frac{56+40}{2} = 48$, $56-48 = 8$. Applying this procedure on all four pairs we obtain the following sequence of numbers:

48	8	16	-8	48	0	28	12

In mathematical terms, we have replaced the pair (a, b) by a new pair (c, d) given by

$$c = \frac{a+b}{2}, \; d = a - \frac{a+b}{2} = \frac{a-b}{2}. \tag{4.10}$$

Let us write the new sequence of numbers under the original sequence:

56	40	8	24	48	48	40	16
48	8	16	-8	48	0	28	12

In some sense, the obtained sequence in the second row contains the same information as the original sequence in the first row: we can always come from the second row and back to the first by inverting the above procedure. That is, we have to solve the equations (4.10) with respect to a, b and apply the obtained formula to the four pairs of numbers in the second row. Solving the equations (4.10) with respect to a, b gives

$$a = c + d, \; b = c - d, \tag{4.11}$$

so the inverse transform is exactly as easy to apply as the transform itself.

When we go back to the original signal from the second row we speak about *reconstruction* of the original information. If our purpose is to store the information or to send it, we can equally well work with the second row as with the first: we just have to remember to transform back at a certain stage. However, at this moment it is not clear that we gain anything by the transformation. We keep this question open for a moment, and apply the procedure once more; but only on the numbers appearing as averages. In other words, we let the numbers $8, -8, 0, 12$ (calculated as differences) stay in the table, and repeat the process on the numbers $48, 16, 48, 28$, i.e., on the pairs $48, 16$ and $48, 28$. That is, we calculate average and difference (in the above sense) for each of these pairs. This leads to the numbers $32, 16, 38, 10$, which are placed on the free places in the table; this gives us the table

32	8	16	-8	38	0	10	12

These eight numbers also represent the original information: we just have to invert the above procedure twice, then we are back at the original sequence. However, some bookkeeping is involved: we have to keep track of which numbers we calculated as averages, and which were differences.

Finally, we repeat the process on the numbers $32, 16, 38, 10$. The numbers $16, 10$ appeared as differences, so we do not change them; the numbers 32,38 appeared as averages, so we replace them by their average 35 and the difference $32 - 35 = -3$. Thus, we obtain the table

35	8	16	-8	-3	0	10	12

So far, we have only argued that the performed operations do not change the information, in the sense that we can come back to the original numbers by repeated application of the inversion formula (4.11). In order to show that we have gained something, we now apply *thresholding* on the numbers in the final table. In popular words, this means that we remove the numbers which numerically are smaller than a certain fixed number; more precisely, we replace them by zeros. If we for example decide to remove all numbers which numerically are smaller than 4, we obtain the table

35	8	16	-8	0	0	10	12

Let us now perform the reconstruction, i.e., apply the inversion formula (4.11), based on these numbers; note that we have to perform the inversion in three steps, paying close attention to which numbers we obtained as averages and differences in each step. Then we obtain the sequence in the first row below; in order to compare with the original sequence we place the signal we started with in the second row:

59	43	11	27	45	45	37	13
56	40	8	24	48	48	40	16

We notice that the numbers in the first row are quite close to the original sequence. If we are rough and use thresholding by 9 in the table obtained after three steps of calculating averages and differences, we obtain the table

35	0	16	0	0	0	10	12

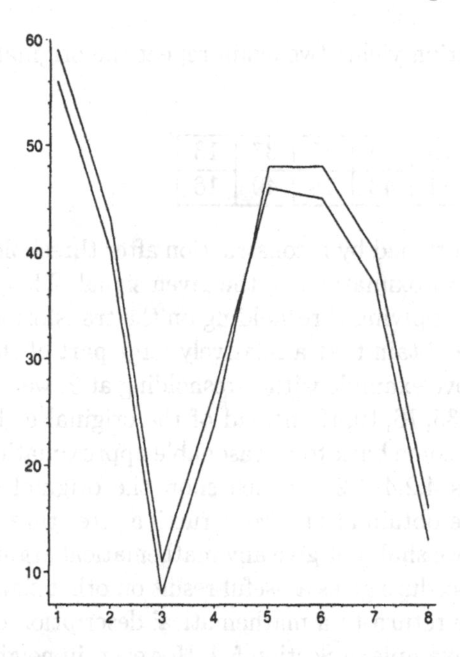

Figure 4.2.4 *The original signal and reconstruction after thresholding with 4.*

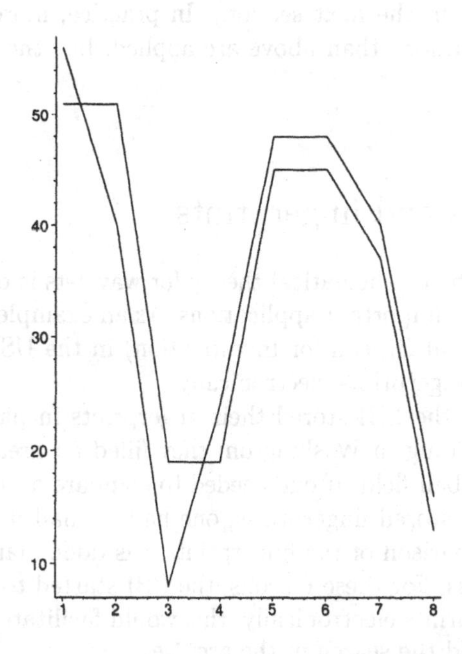

Figure 4.2.5 *The original signal and reconstruction after thresholding with 9. Observe that even though the rough thresholding is worse, the reconstructed signal still follows the shape of the original signal quite well, except in the neighborhood of points where the signal oscillates heavily.*

Here, reconstruction yields (we again repeat the original sequence in the second row)

51	51	19	19	45	45	37	13
56	40	8	24	48	48	40	16

The sequences obtained by reconstruction after thresholding are in many cases acceptable approximations of the given signal. This is the key to the compression: after applying thresholding on the transformed version of the given sequence, we obtain that a relatively large part of the signal consists of zeros: in the above example with thresholding at 9, we only have to store the four numbers $35, 16, 10, 12$ instead of the original eight numbers, and still we are able to come back to a reasonable approximation of the original signal. See Figures 4.2.4–4.2.5, which show the original signal compared with the signals we obtain after reconstruction after thresholding.

At the moment we shall not give any mathematical argument for the fact that the above procedure gives a useful result on other signals than the one in the example; we return to a mathematical description of the operations performed in this example in Section 5.4. However, in neighborhoods where the considered signal does not oscillate too much, it is clear that many differences will be small, and thus by thresholding can be replaced by zeros (see the discussion in the next section). In practice, more advanced and more efficient operations than above are applied, but the basic idea is as sketched. □

4.3 Wavelets and fingerprints

As we have seen, the mathematical theory for wavelets is quite recent, but it has already found important applications. As an example we now discuss how the FBI (Federal Bureau for Investigation) in the USA uses wavelets as a tool to store fingerprints electronically.

For many years, the FBI stored their fingerprints in paper format in a highly secured building in Washington; they filled an area which had the same size as a football field. If one needed to compare a fingerprint in San Francisco with the stored fingerprints, one had to mail it to Washington. Furthermore, comparison of the fingerprints was done manually, so it was a quite slow process. For these reasons, the FBI started to search for ways to store the fingerprints electronically; this would facilitate transmission of the information and the search in the archive.

We can consider a fingerprint as a small picture, so a natural idea is to split each square-inch into, say, 256×256 pixels, to which we associate a gray-tone on a scale from for example 0 (completely white) to 256 (completely black). This way we have kept the essential information in the form

Figure 4.3.1 *Albert Cohen's dog Ram with a small pixel-size.*

Figure 4.3.2 *Ram with a large pixel-size.*

of a sequence of pairs of numbers, namely, the pairs consisting of a numbering of the pixels and the associated gray-tone. The accuracy depends essentially on the pixel-size; see Figures 4.3.1–4.3.2. This sequence can easily be stored and transmitted electronically, i.e., it can if necessary be sent rapidly to San Francisco and be compared with a given fresh fingerprint.

The remaining problem has to do with the size of the archive. With reasonable accuracy, the above procedure will represent each fingerprint by a sequence of numbers which uses about 10 Mb, i.e., information corresponding to about 10 standard diskettes. The FBI has more than 30 million sets of fingerprints (each consisting of 10 fingers) and receives about 30000 new fingerprints each day. Thus, we are speaking about enormous sets of data, and it is necessary to do some compression in order to be able to handle them. This has to be done in a way such that the structure of the fingerprints is kept; see the discussion about compression in Section 2.7.

This is where wavelets enter the scene. The FBI started to search for efficient ways to compress data and became interested in wavelets. It had been known for some time that wavelets are good to represent many types of signals using few coefficients, and the end of the story was that the FBI decided to use a variant of the wavelets discussed here. In fact, the FBI uses Daubechies' biorthogonal spline wavelets; here ψ is piecewise polynomial, and (4.3) is satisfied with coefficients which are a slight modification of (4.5), see Section 5.8. In the concrete case it turned out that it was enough to represent a fingerprint using about 8% of the original information, which makes it possible to store it on a single diskette. The compression method has the impressive name *The Wavelet Scalar Quantization Gray-scale Fingerprint Image Compression Algorithm,* usually abbreviated WSQ.

The fundamental Example 4.2.3 gives part of the explanation why such an efficient compression is possible. A fingerprint (or almost any other picture) has a certain structure, which implies that the sequence of gray-tones associated with the pixels are not random numbers. Consider for example a picture showing a house: here the gray-tones will be almost constant on, e.g., the walls, and only around corners and windows will there be large variations. This can also be observed in Figure 4.3.2; except on the boundary, the dark pillow in the upper left corner is nicely represented even with the large pixel-size, because the gray-tones are almost constant. As described in Example 4.2.3, large areas with almost constant gray-tones (and therefore small differences) give good options for compression, and in the concrete example with fingerprints, excellent results are obtained. See Figure 4.3.3 and Figure 4.3.4, which show an original fingerprint and its compressed version, respectively. The original fingerprint exists electronically and can be downloaded from the NIST (National Institute of Standards and Technology) via the ftp address

ftp://sequoyah.nist.gov/pub/cmp_imgs/cmp_imgs/

Figure 4.3.3 *Original fingerprint.*

Figure 4.3.4 *Fingerprint, compressed using wavelets, and reconstructed.*

Further information can be obtained via

http://www.itl.nist.gov/iad/894.03/fing/cert_gui.html

The compressed fingerprint is made using software developed by the mathematician Mladen Victor Wickerhauser. The software has the name ID♯55291, is owned by the company Wickerhauser Consulting, and can be downloaded from Wickerhauser's home page at Washington University in St.Louis, USA:

http://www.math.wustl.edu/~victor

under "my free software".

Let us finally mention that compression of fingerprints in principle also can be performed using Fourier analysis. However, this classical method is less efficient: at the rate of compression the FBI uses for the wavelet method, it would no longer be possible to follow the contours in a reconstructed fingerprint, and the result would be useless in this special context.

The application of wavelets to compression of fingerprints is only one among several successes of wavelets. In 1986 the research group *JPEG* (Joint Photographic Experts Group) was formed with the purpose of development of international standards for compression of pictures. Also in this context, one of the main goals is to be able to transmit pictures without using too much capacity. We mention that transmission (in good quality) of a color picture consisting of 1024×768 pixels uses about 12 MB. Using a 1GB internet connection transmission of the picture takes 0.1 second; however, if only 64KB is available it takes about 7 minutes. Thus, in this case some compression is needed in order to obtain a useful result. The latest version of the JPEG standard for compression has the name JPEG2000, and uses biorthogonal spline wavelets.

Let us mention one more example where the methods discussed here are important:

Example 4.3.5 In Singapore, a new security system was introduced in Hitachi Tower (a 37-storey office building) in 2003: now, the 1500 employes get access to the building by scanning their fingers. The scanner uses infrared rays to trace the haemoglobin in blood in order to capture the vein patterns in the finger; these patterns determine the person uniquely. After comparing with the scanned data in an electronic archive, it is decided whether the person can get in or not.

In order for a system of this type to work in practice, it is necessary to be able to store the scanned data using as few bits as possible; and it is necessary to have fast algorithms to extract the essential data from the finger being scanned when somebody wants to enter the building. Wavelets fulfil both conditions. □

4.4 Wavelet packets

Let us return to Figure 4.3.2, showing Ram and his pillows. The side-length of the pixels on the figure with the coarse resolution is twice the side-length of the pixels with the fine resolution. Thus, compared to the blurred version, the detailed picture requires that we store and process four times as many coefficients. This is usually not a problem when we deal with a single picture, but it becomes an important issue when dealing with large collections of pictures, or if we want very small pixels.

However, Figure 4.3.2 suggests that we can use the available storage capacity better than just to consider as small pixels as possible. In fact, even the coarse resolution gives a nice presentation of large parts of the picture: for example, except for the contour, the dark pillow in the upper left corner is represented very well, so in this region there is absolutely no reason to consider smaller pixels. A much more efficient method would be to keep the pixel size in such parts of the picture, and only consider finer resolutions in the parts of the picture where it is needed, e.g., at the contour of Ram.

A method for doing so was developed by Wickerhauser and his collaborators around 1992; it has the name *wavelet packets.* It allows us to represent signals via combinations of several wavelet systems, for example two wavelet systems "leading to resolutions on different levels". In this way, it is possible to use the fine resolution only in regions where it is really needed.

Wavelet packets are examples of what is called *adaptive methods.* The reason for this name is that the representation of a given signal depends on the given signal, i.e., the choice of the wavelet system used to represent the information in a given part of the picture depends on the picture. In contrast, standard wavelet analysis uses the same representation, namely (4.3) with the coefficients (4.5) for a fixed function ψ, for all signals.

4.5 Alternatives to wavelets: Gabor systems

Around 1990 many researchers were extremely optimistic about the range of problems wavelets would solve: some researchers almost believed that wavelets would be the right tool for arbitrary problems in science. Since then, a more realistic picture has been obtained: like all other tools, wavelets have their strong sides and their weak sides. For example, it would be wrong to claim that all types of signal transmission should be performed with wavelets: there exist other types of bases, which frequently are better suited for a certain application. In Section 5.9 we give the formal definition of *Gabor systems*, which are also popular in signal analysis. For example, such systems play a central role in several wireless communication systems; e.g., in *Orthogonal Frequency Division Multiplexing,* which is a part of the

European Digital Audio Broadcasting standard. A short introduction to the mathematical properties of Gabor systems is given in Section 5.9.

4.6 Exercises

4.1 Repeat Example 4.2.3 with other choices of the sequence.

4.2 Based on the Haar wavelet, find the coefficients $c_{j,k},\ j \geq 0$ in (4.5) for the function

$$f(x) = \begin{cases} \sin(2\pi x), & x \in [0, 1[, \\ 0, & \text{otherwise.} \end{cases}$$

5

Wavelets and their Mathematical Properties

In this chapter we concentrate on the mathematical properties of wavelets, but still with more weight on an intuitive understanding than on technical details.

Section 5.1 views wavelets as a tool to obtain expansions in Hilbert spaces; in that sense, it is a continuation of Section 3.4. Section 5.2 gives a short introduction to multiresolution analysis, and Section 5.3 explains the role of the Fourier transform in wavelet analysis; it contains examples, showing that one should think about wavelet analysis as an extension of Fourier analysis rather than a modern substitute. Section 5.4 deals with the first wavelet that was ever constructed, namely, the Haar wavelet; in that section, we also introduce some of the wavelet algorithms which make wavelets so useful in practice. Section 5.5 highlights the role of wavelets with compact support, and shows how this property makes the coefficients in the wavelet expansion reflect the properties of the function at hand. Section 5.6 connects to the theme in Section 3.7: it explains how wavelet expansions reflect the local behavior of a function, and how wavelets can be used to detect jumps in a signal. Section 5.7 returns to the subject of best N-term approximation, which we discussed for Fourier series in Section 3.8. Section 5.8 shows how some of the limitations in wavelet theory can be removed by extending the framework; in fact, by giving up the requirement that $\{\psi_{j,k}\}_{j,k\in\mathbb{Z}}$ forms an orthonormal basis for $L^2(\mathbb{R})$ (but still insisting on each function in $L^2(\mathbb{R})$ having a series expansion in terms of $\{\psi_{j,k}\}_{j,k\in\mathbb{Z}}$), we gain much more freedom in the choice of ψ. Finally, Section 5.9 gives a short description of Gabor systems, which are frequently used as alternatives to wavelet systems.

5.1 Wavelets and $L^2(\mathbb{R})$

In Section 3.4 we saw that Fourier series have an exact description in terms of orthonormal bases in the Hilbert space $L^2(-\pi, \pi)$; this way of viewing Fourier series also explains the fact that the Fourier series for a given function does not necessarily converge to the function in the pointwise sense.

In wavelet analysis we are facing exactly the same situation when we want to explain the exact meaning of the representation (4.3) for functions in $L^2(\mathbb{R})$.

If we identify functions which are equal almost everywhere, we can equip $L^2(\mathbb{R})$ with the inner product

$$\langle f, g\rangle = \int_{-\infty}^{\infty} f(x)\overline{g(x)}dx, \quad f, g \in L^2(\mathbb{R}), \tag{5.1}$$

and the associated norm

$$||f|| = \sqrt{\langle f, f\rangle}, \quad f \in L^2(\mathbb{R});$$

then $L^2(\mathbb{R})$ becomes a Hilbert space. The technically correct definition of a wavelet is that this is a function $\psi \in L^2(\mathbb{R})$ for which the functions $\{\psi_{j,k}\}_{j,k\in\mathbb{Z}}$ form an orthonormal basis for $L^2(\mathbb{R})$; translating the abstract definition in (3.16) to the setting of $L^2(\mathbb{R})$, this means that the following two sets of conditions are satisfied:

$$||f||^2 = \sum_{j,k\in\mathbb{Z}} |\langle f, \psi_{j,k}\rangle|^2, \quad \forall f \in L^2(\mathbb{R}), \tag{5.2}$$

$$\langle \psi_{j,k}, \psi_{j',k'}\rangle = \begin{cases} 1 \text{ if } k = k', j = j', \\ 0 \text{ otherwise.} \end{cases} \tag{5.3}$$

By (3.17) we know that this implies that each $f \in L^2(\mathbb{R})$ has the representation

$$f = \sum_{j,k\in\mathbb{Z}} \langle f, \psi_{j,k}\rangle \psi_{j,k},$$

understood in the sense that

$$\left|\left| f - \sum_{|j|,|k|\leq N} \langle f, \psi_{j,k}\rangle \psi_{j,k} \right|\right| \to 0 \text{ as } N \to \infty. \tag{5.4}$$

Note that the sum in (5.4) uses a special indexing of the elements in the wavelet system. This is not crucial: one can prove that the wavelet expansion converges unconditionally in $L^2(\mathbb{R})$.

It is clear from the definition that only very special functions ψ can be wavelets. However, there exists a general framework for construction of such functions; this is the subject for the next section.

5.2 Multiresolution analysis

We already mentioned that multiresolution analysis is a central ingredient in wavelet analysis. Here is the definition:

Definition 5.2.1 *A multiresolution analysis consists of a sequence* $\{V_j\}_{j\in\mathbb{Z}}$ *of closed subspaces of* $L^2(\mathbb{R})$ *and a function* $\phi \in V_0$*, satisfying the conditions*

(i) $\cdots V_{-1} \subset V_0 \subset V_1 \cdots$;

(ii) $\overline{\cup_j V_j} = L^2(\mathbb{R})$ *and* $\cap_j V_j = \{0\}$;

(iii) $f \in V_j \Leftrightarrow [x \mapsto f(2x)] \in V_{j+1}$;

(iv) $f \in V_0 \Rightarrow [x \mapsto f(x-k)] \in V_0, \ \forall k \in \mathbb{Z}$;

(v) $\{\phi(\cdot - k)\}_{k\in\mathbb{Z}}$ *is an orthonormal basis for* V_0.

In connection with approximation theory, approximation of a function in $L^2(\mathbb{R})$ will be performed via a function from one of the spaces V_j. The importance of multiresolution analysis partly lies in the fact that the conditions guarantee the existence of convenient transforms between the spaces V_j. These transformations appear in a special case in Section 5.4.

Multiresolution analysis (abbreviated MRA) is the main subject in the majority of the published wavelet books. We will not discuss it further in this book, but only mention its relationship with the construction of wavelet bases.

The assumptions that we made in Definition 5.2.1 imply that the functions $\{2^{1/2}\psi(2x-k)\}_{k\in\mathbb{Z}}$ form an orthonormal basis for V_1. Since $\phi \in V_0 \subset V_1$, this implies that there exist coefficients $\{c_k\}_{k\in\mathbb{Z}} \in \ell^2$ such that

$$\phi(x) = \sum_{k\in\mathbb{Z}} c_k \phi(2x-k).$$

Now, it can be proved that the function

$$\psi(x) := \sum_{k\in\mathbb{Z}} (-1)^k \overline{c_{1-k}} \phi(2x-k)$$

generates an orthonormal basis $\{\psi_{j,k}\}_{j,k\in\mathbb{Z}}$ for $L^2(\mathbb{R})$.

Without proof, we mention that the Haar wavelet can be constructed via an MRA.

5.3 The role of the Fourier transform

Wavelet analysis is often looked at as some kind of modern Fourier analysis. The fact that wavelet analysis has been so successful in many applications can make the reader wonder whether it eventually will replace Fourier analysis. The answer is no. In fact, a more technical description of wavelets will show that classical Fourier analysis is a very important ingredient in wavelet theory. We will make this point clear by considering some places in wavelet analysis where Fourier analysis plays an important role.

Example 5.3.1 We already mentioned that multiresolution analysis is a tool to construct wavelets, which in many instances is very efficient in order to find wavelets with prescribed properties. However, a common feature of almost all MRA-based constructions presented in the literature so far is that they do not directly lead to a function ψ which is a wavelet; rather, they lead to a construction of a function ν such that the *inverse Fourier transform* of ν is a wavelet. That is, the wavelet ψ has to be found via the equation $\hat{\psi} = \nu$; alternatively, via our expression (3.37) for the inverse Fourier transform,

$$\psi(x) = \mathcal{F}^{-1}\nu(x) = \int_{-\infty}^{\infty} \nu(\gamma)e^{2\pi i x \gamma}d\gamma.$$

Thus, classical Fourier analysis is an integrated part of the procedure. □

Example 5.3.2 Let us mention one more example related to MRA. One of the conditions in Definition 5.2.1 says that we have to select a function ϕ for which

$$\int_{-\infty}^{\infty} \phi(x-k)\overline{\phi(x-k')}dx = \begin{cases} 1 \text{ if } k = k', \\ 0 \text{ otherwise.} \end{cases} \tag{5.5}$$

Observe that this condition is part of the condition (4.4) as well. In practice, a function ϕ satisfying (5.5) is often found in terms of its Fourier transform; in fact, one can prove that (5.5) is equivalent to the condition

$$\sum_{k \in \mathbb{Z}} \left|\hat{\phi}(\gamma + k)\right|^2 = 1, \ \ \gamma \in \mathbb{R}.$$ □

Example 5.3.3 Historically, the first step in the modern development of wavelets was to consider an integral transformation, the so-called *wavelet transform*. Given a function $\psi \in L^2(\mathbb{R})$ satisfying a certain *admissibility condition*, Grossmann et. al. [14] defined the wavelet transformation of a function $f \in L^2(\mathbb{R})$ by

$$W_\psi(f)(a,b) = \int_{-\infty}^{\infty} f(x)\frac{1}{|a|^{1/2}}\overline{\psi(\frac{x-b}{a})}dx, \ a \neq 0, b \in \mathbb{R}.$$

The wavelet transformation leads to an integral representation of functions in $L^2(\mathbb{R})$; in fact, letting

$$\psi^{a,b}(x) = \frac{1}{|a|^{1/2}}\overline{\psi(\frac{x-b}{a})},\ a \neq 0, b \in \mathbb{R}$$

one can prove that

$$f(x) = \int_{-\infty}^{\infty}\int_{-\infty}^{\infty} W_\psi(f)(a,b)\psi^{a,b}(x)\frac{dadb}{a^2},\ f \in L^2(\mathbb{R})$$

(we shall not go into details with the exact meaning of the integral). The admissibility condition needed to make this work is most conveniently expressed via the Fourier transform, in form of the condition

$$\int_{-\infty}^{\infty}\frac{|\hat{\psi}(\gamma)|^2}{|\gamma|}d\gamma < \infty.$$

□

5.4 The Haar wavelet

We already mentioned that the first wavelet ever constructed was the Haar wavelet,

$$\psi(x) = \begin{cases} 1, & x \in [0, \frac{1}{2}[, \\ -1, & x \in [\frac{1}{2}, 1[, \\ 0, & \text{otherwise.} \end{cases} \tag{5.6}$$

In the entire section, ψ will denote this particular wavelet. We will call this function as well as the scaled and translated versions $\psi_{j,k}$ for *Haar functions.*

In the entire section we consider a continuous function

$$f : [0,1] \to \mathbb{R}.$$

Our aim is to approximate such a function via piecewise constant functions. As we will see, the Haar wavelet shows up in our attempt to do so. Our final goal is to prove Theorem 5.4.13 and Theorem 5.4.14.

A natural way to approximate f is to split $[0,1]$ into a certain number of intervals, and then, on each of these intervals, approximate $f(x)$ by the average of f over the interval. We will focus on the case where we split $[0,1]$ into 2^k intervals for some $k = 0,1,\ldots$; we will further choose the intervals to have equal length. Ignoring the endpoint $x = 1$, this means that our intervals are

$$I_n = [n2^{-k}, (n+1)2^{-k}[,\ n = 0,1,\ldots,2^k - 1.$$

These intervals have length 2^{-k}, so the average of f on I_n is

$$a_n = 2^k \int_{n2^{-k}}^{(n+1)2^{-k}} f(x)dx. \tag{5.7}$$

Note that the intervals I_n as well as the coefficients a_n actually depend on the chosen value of k. Thus, it would have been more accurate to denote the intervals by $I_{k,n}$ and the coefficients by $a_{k,n}$; however, except in Theorem 5.4.13 we will suppress the dependence on k.

If k is sufficiently large (i.e., if the intervals I_n are sufficiently small), the value a_n is a reasonable approximation to $f(x)$ for $x \in I_n$; thus, as an approximation of f on $[0,1]$ it is natural to take

$$f_k(x) := \sum_{n=0}^{2^k-1} a_n \chi_{I_n}(x) = \sum_{n=0}^{2^k-1} a_n \chi_{[n2^{-k},(n+1)2^{-k}[}(x). \tag{5.8}$$

The parameter k indicates the *level,* or *scale,* of the approximation: large values of k lead to fine approximations, while small values give us coarse approximations which are constant on large intervals. By choosing k sufficiently large, we can obtain that f_k approximates f as well as we would like in the uniform sense:

Lemma 5.4.1 *Assume that $f : [0,1] \to \mathbb{R}$ is continuous. Then, for any $\epsilon > 0$ there exists $K \in \mathbb{N}$ such that*

$$|f(x) - f_k(x)| \leq \epsilon, \; \forall x \in [0,1[,$$

for all $k \geq K$.

Proof: A continuous function on a bounded and closed interval is uniformly continuous, i.e., for any given $\epsilon > 0$ we can choose $\delta > 0$ such that

$$|x-y| \leq \delta, \; x,y \in [0,1] \Rightarrow |f(x) - f(y)| \leq \epsilon.$$

If we now choose $K \in \mathbb{N}$ such that $2^{-K} \leq \delta$, then $2^{-k} \leq \delta$ for all $k \geq K$. Thus, for such k, the maximal variation of $f(x)$ on each interval I_n is at most ϵ. Now, considering an arbitrary $x \in [0,1[$, this point belongs to exactly one of the intervals, say, I_n, and therefore

$$|f(x) - f_k(x)| = |f(x) - a_n|.$$

Since a_n is an average of function values which deviate at most ϵ from $f(x)$, we obtain the announced result. $\square$

Let us write the first few approximations in (5.8) explicitly:

$$f_0(x) = a\chi_{[0,1[}(x), \text{ where } a = \int_0^1 f(x)dx, \tag{5.9}$$

$$f_1(x) = \sum_{n=0}^{1} b_n\chi_{[n/2,(n+1)/2[}(x), \text{ where } b_n = 2\int_{n/2}^{(n+1)/2} f(x)dx, \tag{5.10}$$

$$f_2(x) = \sum_{n=0}^{3} c_n\chi_{[n/4,(n+1)/4[}(x), \text{ where } c_n = 4\int_{n/4}^{(n+1)/4} f(x)dx, \tag{5.11}$$

$$f_3(x) = \sum_{n=0}^{7} d_n\chi_{[n/8,(n+1)/8[}(x), \text{ where } d_n = 8\int_{n/8}^{(n+1)/8} f(x)dx. \tag{5.12}$$

Figures 5.4.6–5.4.3 show these approximations for a concrete function f. We see that larger values of k lead to better approximations; however, when we increase k, we also increase the number of coefficients appearing in the expansion of f_k, i.e., we increase the effort we need to spend on calculation and storage. For a given value of k, the approximation f_k involves 2^k coefficients, so the effort increases exponentially as a function of k.

As we see in Figure 5.4.6, for this concrete function $k = 3$ does not lead to a good approximation: much larger values are usually needed. This means that good approximations might require very large data sets. Therefore it becomes important to be able to compress the data set, or at least know how to represent the essential information via fewer coefficients. A key to this type of results turns out to be the existence of a nice relationship between the coefficients associated to approximation with a certain value of k, and the coefficients on the "coarser level" $k - 1$. In the next example we consider this relationship for $k = 2$.

Example 5.4.2 Let us compare the approximations f_1 and f_2. We will refer to the notation used in (5.10) and (5.11). Thinking of f_1 as a "blurred" version of f_2, it is natural to write

$$f_2(x) = f_1(x) + \text{ details} = f_1(x) + [f_2(x) - f_1(x)].$$

For each given $x \in [0,1[$, $f_1(x)$ is the average of f_2 over two neighbor intervals, so the function $f_2 - f_1$ is a linear combination of the Haar function $\psi(2\cdot)$ and its translate $\psi(2\cdot -1)$. See Figure 5.4.9; in fact, the representation of f_2 in terms of f_1 is given by

$$f_2(x) = \sum_{n=0}^{3} c_n\chi_{[n/4,(n+1)/4[}(x) \tag{5.13}$$

$$= f_1(x) + \text{ details}$$

$$= \sum_{n=0}^{1} b_n\chi_{[n/2,(n+1)/2[}(x) + \sum_{n=0}^{1} s_n\psi(2x-n), \tag{5.14}$$

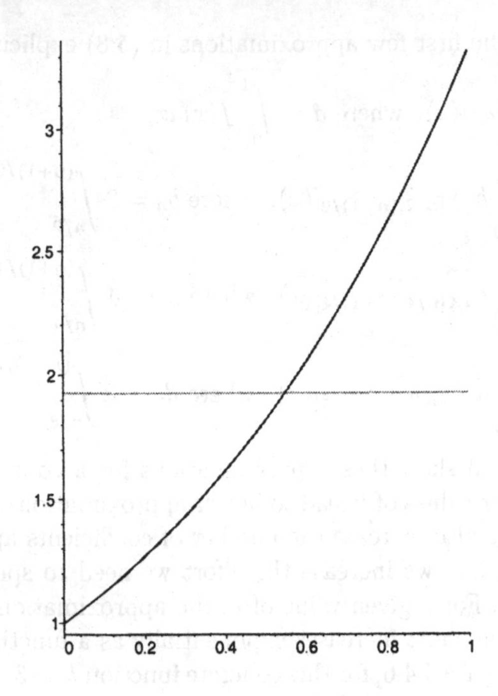

Figure 5.4.3 *A function f and its approximation f_0.*

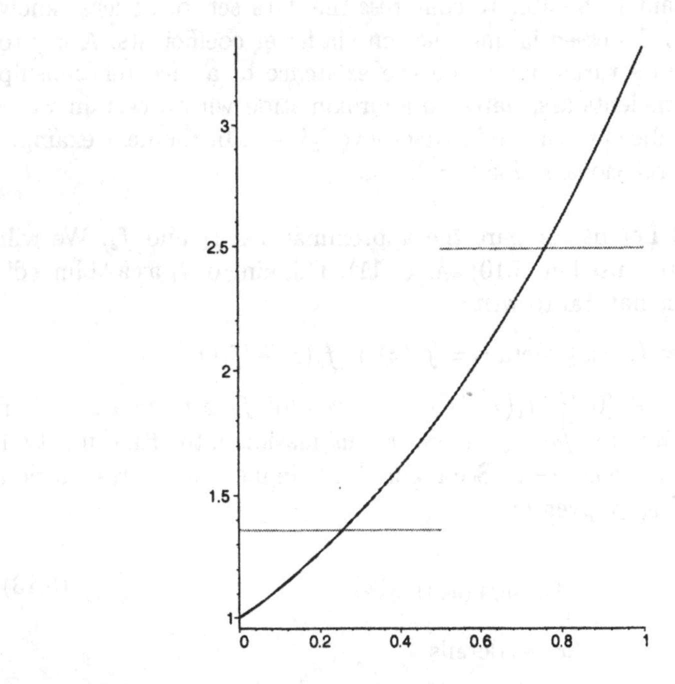

Figure 5.4.4 *A function f and its approximation f_1.*

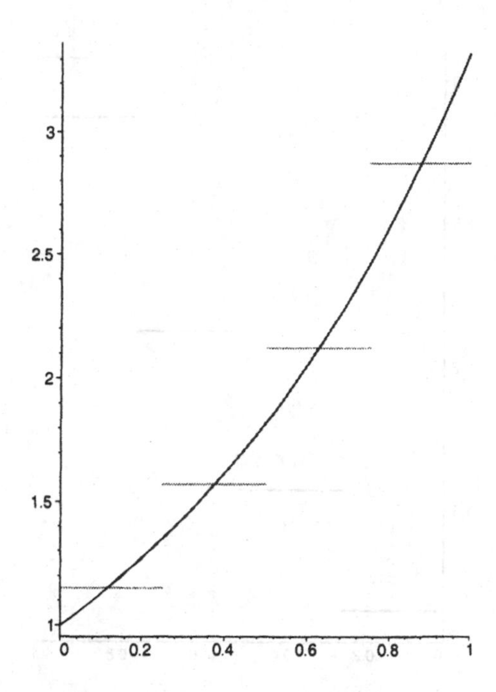

Figure 5.4.5 *A function f and its approximation f_2.*

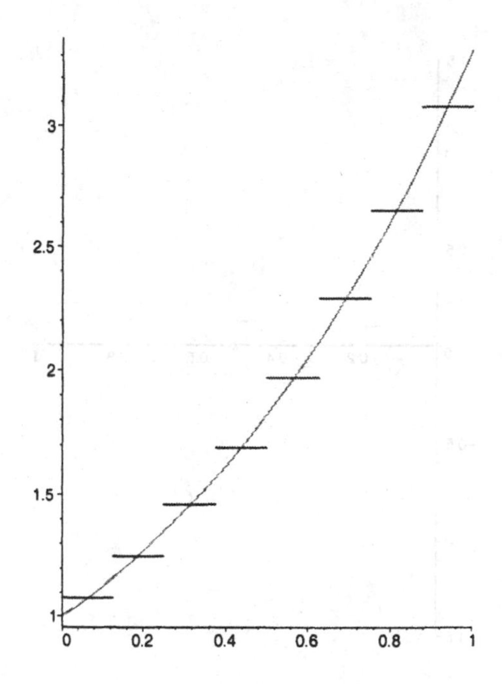

Figure 5.4.6 *A function f and its approximation f_3.*

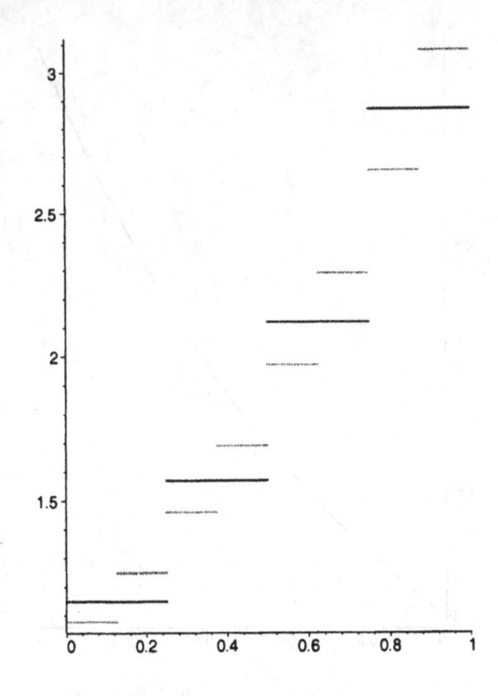

Figure 5.4.7 *The functions f_3 and f_2.*

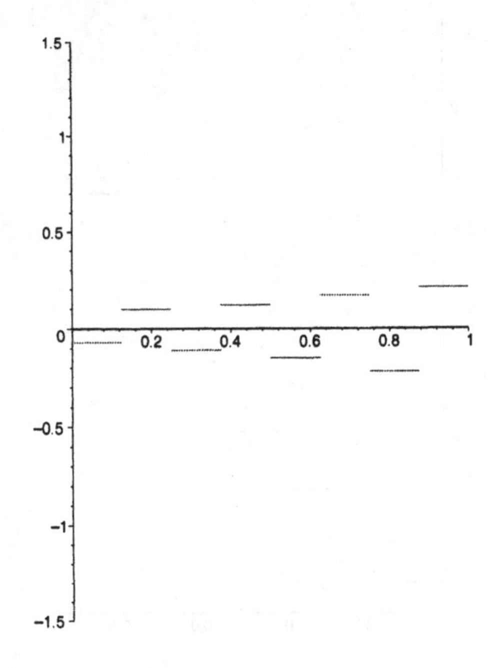

Figure 5.4.8 *The function $f_3 - f_2$.*

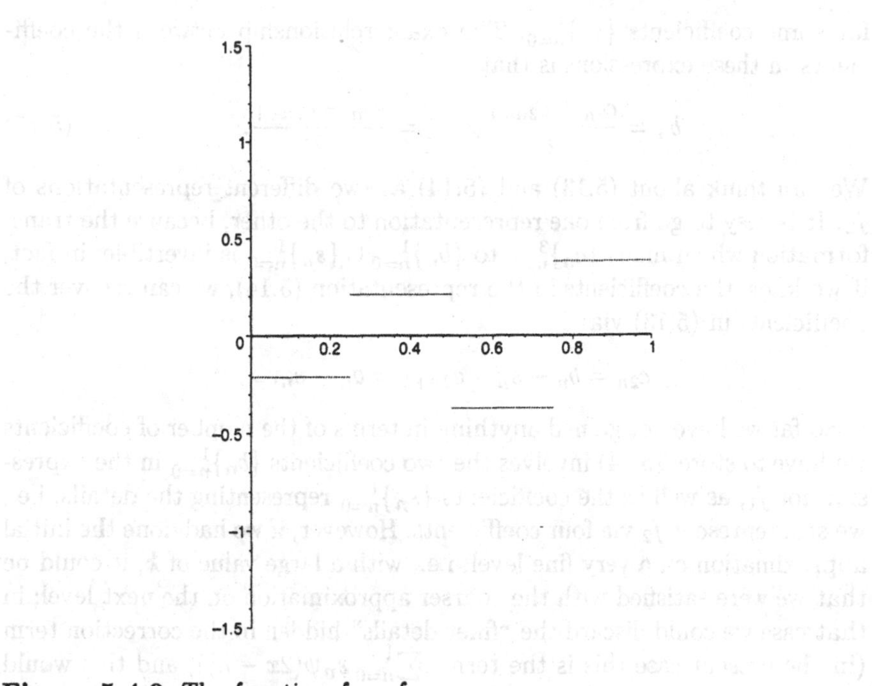

Figure 5.4.9 *The function* $f_2 - f_1$.

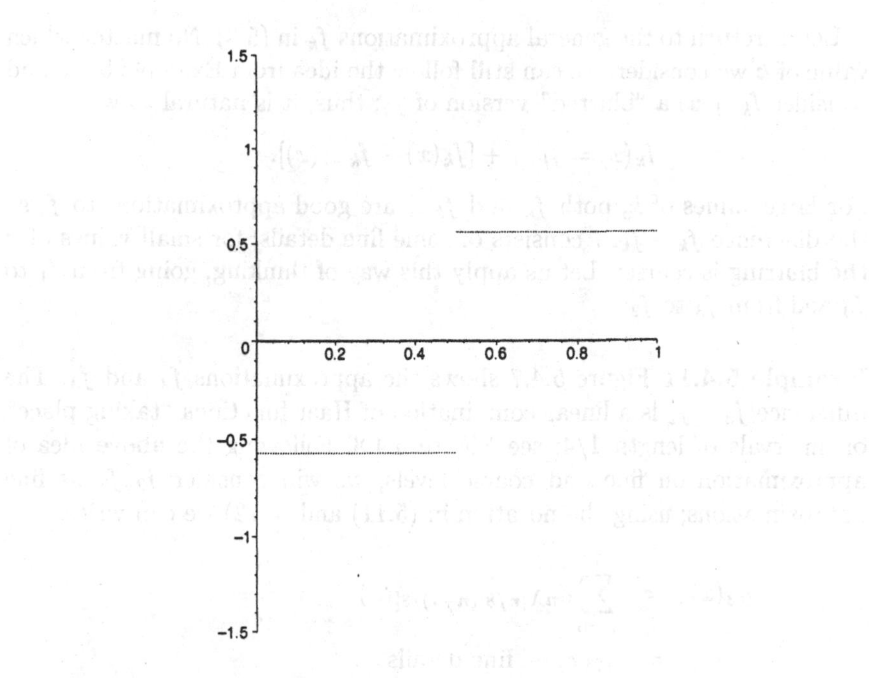

Figure 5.4.10 *The function* $f_1 - f_0$.

for some coefficients $\{s_n\}_{n=0}^{1}$. The exact relationship between the coefficients in these expressions is that

$$b_n = \frac{c_{2n} + c_{2n+1}}{2}, \quad s_n = \frac{c_{2n} - c_{2n+1}}{2}. \tag{5.15}$$

We can think about (5.13) and (5.14) as two different representations of f_2. It is easy to go from one representation to the other, because the transformation which maps $\{c_n\}_{n=0}^{3}$ to $\{b_n\}_{n=0}^{1} \cup \{s_n\}_{n=0}^{1}$ is invertible; in fact, if we know the coefficients in the representation (5.14), we can recover the coefficients in (5.13) via

$$c_{2n} = b_n + s_n, \quad c_{2n+1} = b_n - s_n.$$

So far we have not gained anything in terms of the number of coefficients we have to store: (5.14) involves the two coefficients $\{b_n\}_{n=0}^{1}$ in the expression for f_1, as well as the coefficients $\{s_n\}_{n=0}^{1}$ representing the details, i.e., we still represent f_2 via four coefficients. However, if we had done the initial approximation on a very fine level, i.e., with a large value of k, it could be that we were satisfied with the coarser approximation on the next level; in that case we could discard the "finer details" hidden in the correction term (in the present case this is the term $\sum_{n=0}^{1} s_n\psi(2x-n)$), and this would reduce the number of coefficients to store or process by a factor of two. □

Let us return to the general approximations f_k in (5.8). No matter which value of k we consider, we can still follow the idea from Example 5.4.2 and consider f_{k-1} as a "blurred" version of f_k; thus, it is natural to write

$$f_k(x) = f_{k-1} + [f_k(x) - f_{k-1}(x)].$$

For large values of k, both f_k and f_{k-1} are good approximations to f, so the difference $f_k - f_{k-1}$ consists of some fine details; for small values of k the blurring is coarser. Let us apply this way of thinking, going from f_1 to f_0 and from f_3 to f_2:

Example 5.4.11 Figure 5.4.7 shows the approximations f_2 and f_3. The difference $f_3 - f_2$ is a linear combination of Haar functions "taking place" on intervals of length $1/4$; see Figure 5.4.8. Following the above idea of approximation on fine and coarse levels, we will consider f_2, f_3 as fine approximations; using the notation in (5.11) and (5.12) we can write

$$\begin{aligned} f_3(x) &= \sum_{n=0}^{7} d_n \chi_{[n/8,(n+1)/8[}(x) \\ &= f_2(x) + \text{ fine details} \\ &= \sum_{n=0}^{3} c_n \chi_{[n/4,(n+1)/4[}(x) + \sum_{n=0}^{3} r_n \psi(2^2x - n). \end{aligned}$$

The relationship between the coefficients on the finest level, i.e., the coefficients in f_3, and the coefficients in f_2 plus the fine details, is exactly the same as in Example 5.4.2:

$$c_n = \frac{d_{2n} + d_{2n+1}}{2}, \quad r_n = \frac{d_{2n} - d_{2n+1}}{2}. \tag{5.16}$$

Let us finally do the same for the very coarse approximations f_0 and f_1. In this case, the difference $f_1 - f_0$ is a Haar function on the interval $[0, 1]$, see Figure 5.4.10. Again, via (5.9) and (5.10),

$$\begin{aligned} f_1(x) &= \sum_{n=0}^{1} b_n \chi_{[n/2,(n+1)/2[}(x) \\ &= f_0(x) + \text{ coarse details} \\ &= a\chi_{[0,1]}(x) + t\psi(x). \end{aligned}$$

Here we have

$$a = \frac{b_0 + b_1}{2}, \quad t = \frac{b_0 - b_1}{2}. \tag{5.17}$$

□

So far we have transformed between one level and the next. An interesting decomposition is obtained if we apply several steps of this procedure:

Example 5.4.12 Let us again consider f_3 as the finest approximation. Combining what we already stated in Example 5.4.2 and Example 5.4.11, we can write f_3 as a sum of the most blurred approximation f_0 and details on finer and finer scales; in a mixture of symbols and words, the idea is to write

$$\begin{aligned} f_3(x) &= f_2(x) + \text{fine details} \\ &= f_1(x) + \text{details} + \text{fine details} \\ &= f_0(x) + \text{ coarse details} + \text{details} + \text{fine details}. \end{aligned}$$

On a more formal level, this means that

$$\begin{aligned} f_3(x) &= \sum_{n=0}^{7} d_n \chi_{[n/8,(n+1)/8[}(x) \qquad (5.18) \\ &= f_2(x) + \sum_{n=0}^{3} r_n \psi(2^2 x - n) \\ &= f_1(x) + \sum_{n=0}^{1} s_n \psi(2x - n) + \sum_{n=0}^{3} r_n \psi(2^2 x - n) \\ &= a\chi_{[0,1[}(x) + t\psi(x) + \sum_{n=0}^{1} s_n \psi(2x - n) + \sum_{n=0}^{3} r_n \psi(2^2 x - n). \end{aligned}$$

Thus, we can write

$$f_3(x) = a\chi_{[0,1[}(x) + \sum_{j=0}^{2}\sum_{n=0}^{2^j-1} d_{j,n}\psi(2^jx-n), \tag{5.19}$$

where

$$d_{j,n} = \begin{cases} t \text{ for } j=0, n=0, \\ s_n \text{ for } j=1, n=0,1 \\ r_n \text{ for } j=2, n=0,1,2,3. \end{cases}$$

From the point of view of expressing our information using as few coefficients as possible, we have not gained anything so far: (5.18) as well as (5.19) require eight coefficients in order to determine the function f_3. However, conceptually we gain insight about f_3 via (5.19): this expression shows that we can consider f_3 as composed of a "basic approximation" f_0 plus finer and finer details. The formula (5.19) is what we will call a *multiscale representation* of f_3. □

Before we mention some of the advantages of a multiscale representation, we note that all that has been said so far can be done much more generally: exactly the same considerations reveal the relationship between f_k and f_{k-1} for any value of $k \in \mathbb{N}$, and iteration leads to a multiscale representation of f_k. Let us state the result formally. In this description we will make the dependence on k explicit by denoting the coefficients (5.7) appearing in the expansion (5.8) of f_k by $a_{k,n}$ instead of just a_n; that is,

$$a_{k,n} = 2^k \int_{n2^{-k}}^{(n+1)2^{-k}} f(x)dx. \tag{5.20}$$

Theorem 5.4.13 *Let $f : [0,1] \to \mathbb{R}$ be a continuous function. Then, for any value of $k \in \mathbb{N}$, the relationship between the approximations f_k and f_{k-1} is given by*

$$\begin{aligned} f_k(x) &= \sum_{n=0}^{2^k-1} a_{k,n}\chi_{[n2^{-k},(n+1)2^{-k}[}(x) \\ &= f_{k-1}(x) + \textit{ details} \\ &= \sum_{n=0}^{2^{k-1}-1} a_{k-1,n}\chi_{[n2^{-k+1},(n+1)2^{-k+1}[}(x) \\ &\quad + \sum_{n=0}^{2^{k-1}-1} d_{k-1,n}\psi(2^{k-1}x-n), \end{aligned}$$

where the coefficients are related via

$$a_{k-1,n} = \frac{a_{k,2n} + a_{k,2n+1}}{2}, \quad d_{k-1,n} = \frac{a_{k,2n} - a_{k,2n+1}}{2}. \tag{5.21}$$

Furthermore, f_k has the multiscale representation

$$f_k(x) = a\chi_{[0,1[}(x) + \sum_{j=0}^{k-1} \sum_{n=0}^{2^j-1} d_{j,n}\psi(2^j x - n). \tag{5.22}$$

Note that the transformation (5.21) between the coefficients on different levels is exactly the one that appeared in Example 4.2.3 without further explanation.

Formula (5.22) looks very technical; however, it is just the formal expression for the stepwise approximations in

$$\begin{aligned} f_k &= f_{k-1} + \text{ fine details} \\ &= f_{k-2} + \text{ details} + \text{fine details} \\ &= \cdots\cdots \\ &= f_0 + \text{very coarse details} + \text{coarse details} \\ &\quad + \cdots + \text{fine details}. \end{aligned}$$

Let us discuss some of the advantages of the obtained multiscale representations. We will focus on the following main points, which will be explained in more detail below:

- The decomposition (5.22) is *hierarchic:* it decomposes f_k into a basic function (here $f_0 = a\chi_{[0,1[}$) plus details on finer and finer levels.
- Given any value of k, the coefficients on the level $k-1$ and the coefficients appearing in the details can be found from the coefficient on level k via (5.21); see (5.15), (5.16) and (5.17) for the first few cases.
- The transformation (5.21) between coefficients on one level and the pairs

  (coefficients on the coarser level, details)

  is invertible.
- The information in (5.22) is *local:* for a given $x \in [0,1]$, we only get contributions in the details for the values of j, n for which $2^j x - n \in [0,1[$; that is, given j, we only have contributions for the n-values for which $x \in [2^{-j}n, 2^{-j}(n+1)[$.
- The representation (5.22) contains *frequency information* about f: slow oscillations in f will lead to nonzero coefficients for small values of j, while fast oscillations will lead to nonzero coefficients for large values of j (provided k is large).

We will now discuss these issues in more detail, and relate them to the more informal presentation of wavelet theory in Chapter 4.

The hierarchic structure of the multiscale representation makes it possible to get a rough idea about the function f using relatively few coefficients: looking at the approximation f_k at a certain level, we can simply discard some of the finer details from the representation in (5.22), i.e., remove some of the terms in the details $\sum_{n=0}^{2^j-1} d_{j,n}\psi(2^j x - n)$ for large values of j.

In terms of the number of coefficients we need to calculate f_k, the representation (5.22) does not immediately help: we still represent f_k using 2^k coefficients. However, if we are satisfied with a less exact approximation of f, we can discard some of the coefficients appearing in the finer details; and the *locality* means that we can decide in which parts of the signals we want the fine resolution, and where a coarser resolution is satisfying. All together, this allows us to *compress* the information: we will be able to deliver a good approximation of f_k using fewer coefficients.

We already noticed that the transformation (5.21) between the coefficients in f_k, respectively, f_{k-1} and the details, is the same for all values of k; this fact makes it very easy to move around between representations on different levels. In neighborhoods where the considered signal varies slowly, a few applications of this transform (starting with a certain f_k) will often lead to a sequence where many numbers are small. This is the point where *thresholding* comes in: in order to compress the information we will often decide to discard these terms by replacing them by zeroes. Now, using that the transform between the levels is invertible, we can use the inverse transform a number of times and come back to a function on the initial level, which is a good approximation of the original signal. This is exactly what we did in Example 4.2.3.

The fact that oscillations with different frequencies appear with different values of j in (5.22) is very useful in time-frequency analysis. In fact, combined with the locality of the representation, it gives us an idea about the frequencies which appear at a certain time. Furthermore, this aspect is the key to *noise reduction*. Think about, e.g., a tape containing an old recording of a speech or a piano piece: here, most of the noise usually consists of frequencies which are higher than the one appearing in the signal itself. Now, if we simply remove the coefficients in (5.22) corresponding to the high frequencies, we obtain a signal which is cleaned for at least some of the appearing noise; on page 93, we discussed a case where this has been used very effectively on a recording from 1889.

We note that the sums appearing in the expression (5.22) for the functions f_k are finite; however, the number of terms in the series increase with k, i.e., with the desired precision of the approximation. If we aim at an *exact* representation of the given function f, we end up with an infinite series. In our formulation of this result we use the inner product in (5.1):

Theorem 5.4.14 *Assume that* $f : [0,1] \to \mathbb{R}$ *is continuous. Then*

$$f(x) = \langle f, \chi_{[0,1[} \rangle \chi_{[0,1[}(x) + \sum_{j=0}^{\infty} \sum_{n=0}^{2^j-1} \langle f, \psi_{j,n} \rangle \psi_{j,n}(x). \tag{5.23}$$

Proof: Let us first calculate the coefficients $d_{j,k}$ in the multiscale representation of f_k in (5.22). Via (5.21) and (5.20),

$$\begin{aligned} d_{j,n} &= \frac{a_{j+1,2n} - a_{j+1,2n+1}}{2} \\ &= \frac{2^{j+1}}{2} \left(\int_{2n2^{-(j+1)}}^{(2n+1)2^{-(j+1)}} f(x)dx - \int_{(2n+1)2^{-(j+1)}}^{(2n+2)2^{-(j+1)}} f(x)dx \right) \\ &= 2^{j/2} 2^{j/2} \left(\int_{n2^{-j}}^{(n+1/2)2^{-j}} f(x)dx - \int_{(n+1/2)2^{-j}}^{(n+1)2^{-j}} f(x)dx \right) \\ &= 2^{j/2} \int_{n2^{-j}}^{(n+1)2^{-j}} f(x)\psi_{j,n}(x)dx \\ &= 2^{j/2} \langle f, \psi_{j,n} \rangle. \end{aligned}$$

Now, according to Lemma 5.4.1,

$$\begin{aligned} f(x) &= \lim_{k\to\infty} f_k(x) \\ &= a\chi_{[0,1[}(x) + \sum_{j=0}^{\infty} \sum_{n=0}^{2^j-1} d_{j,n} \psi(2^j x - n) \\ &= a\chi_{[0,1[}(x) + \sum_{j=0}^{\infty} \sum_{n=0}^{2^j-1} 2^{j/2} \langle f, \psi_{j,n} \rangle \psi(2^j x - n) \\ &= a\chi_{[0,1[}(x) + \sum_{j=0}^{\infty} \sum_{n=0}^{2^j-1} \langle f, \psi_{j,n} \rangle \psi_{j,n}(x). \end{aligned}$$

□

Let us finally relate the approximations f_k to observations from daily life. We first note that transformation from one level to another is very similar to what happens when we are moving towards an object we look at: at the begining we only see the rough contour, but getting closer and closer, we see more and more details. This corresponds to going from an approximation f_k with a low value of k to approximations with higher values of k.

We already mentioned that a picture is described via a two-dimensional version of the theory discussed so far. Consider a picture, which we think about as being represented by a function f_k. Looking at, e.g., Figures 4.3.1–4.3.2, the locality of the wavelet representation means that we have the option to use many coefficients around the contour of the dog, such that we get a sharp resolution here; and less coefficients on the inside of the pillows, where a coarser resolution still gives a good image.

5.5 The role of compact support

"Real life" techniques can not deal with signals going on "forever": in order to be able to process a signal, it needs to have a finite duration in time. The same restriction applies to all functions appearing in wavelet analysis, and in practice this means that we would like our wavelet ψ to have compact support. So far we have only introduced the terminology "compact support" in the sense of a function which is zero outside a closed and bounded interval in $\mathbb{R}$. The exact meaning of the word *support* for a function f is that

$$\text{supp} f := \overline{\{x \in \mathbb{R} \mid f(x) \neq 0\}},$$

where the bar means closure. So, essentially the support is the set of x for which $f(x) \neq 0$, but with some end points added in order to obtain a closed set.

Compact support of ψ leads to another nice feature of wavelet representations, which is different from what we know from Fourier analysis. Assume that a wavelet ψ has support on the interval $[0, M]$ for some $M \in \mathbb{N}$; then

$$\text{supp } \psi_{j,k} = [2^{-j}k, 2^{-j}(k+M)].$$

Given a signal f, let us now return to the fundamental representation in (4.3), namely,

$$f(x) = \sum_{j\in\mathbb{Z}} \sum_{k\in\mathbb{Z}} c_{j,k}\psi_{j,k}(x) \quad \text{with} \quad c_{j,k} = \int_{-\infty}^{\infty} f(x)\overline{\psi_{j,k}(x)}dx. \tag{5.24}$$

Our knowledge of the support of $\psi_{j,k}$ implies that

$$c_{j,k} = \int_{2^{-j}k}^{2^{-j}(k+M)} f(x)\overline{\psi_{j,k}(x)}dx;$$

that is, the coefficient $c_{j,k}$ only depends on the behavior of f on the interval $[2^{-j}k, 2^{-j}(k+M)]$. In other words, the coefficients $c_{j,k}$ contain information about the analyzed function f *locally*. If we change the function f on a small interval, say $[0, 1]$, then most of the coefficients $c_{j,k}$ remain unchanged; we only obtain new coefficients for the values $j, k \in \mathbb{Z}$ for which

$$[0,1] \cap [2^{-j}k, 2^{-j}(k+M)] \neq \emptyset.$$

This property is very different from what we obtain via Fourier analysis. For example, the Fourier transform of a function f is defined as an integral over $\mathbb{R}$; in general, no matter how little we change the function, this implies that all function values of $\hat{f}$ will change. The same is the case with the Fourier coefficients for a periodic function: an arbitrarily small perturbation of a function in general changes all the Fourier coefficients.

5.6 Wavelets and singularities

Wavelets are excellent to analyze functions in many contexts. To be more exact, if ψ is a wavelet, then important information about a given function f can be extracted from knowledge of the coefficients

$$c_{j,k} = \int_{-\infty}^{\infty} f(x)\overline{\psi_{j,k}(x)}dx \tag{5.25}$$

in (5.24).

An example of this is given in [30], where Walnut shows how a discontinuity of a function f at a certain point is reflected in the coefficients $\{c_{j,k}\}$:

Example 5.6.1 Let again ψ be the Haar wavelet. Let $x_0 \in]0,1[$ and assume that $f : [0,1] \to \mathbb{C}$ is a function which is twice differentiable on the intervals $[0, x_0[$ and $]x_0, 1]$; we also assume that the limits

$$f(x_0^+) := \lim_{x \to x_0^+} f(x) \text{ and } f(x_0^-) := \lim_{x \to x_0^-} f(x)$$

exist. In the following we think about f as a function defined on $\mathbb{R}$, simply by defining $f(x) = 0$ for $x \notin [0,1]$. We now consider the coefficients $c_{j,k}$ in (5.25): since the support of $\psi_{j,k}$ is the interval $I_{j,k} = [2^{-j}k, 2^{-j}(k+1)]$, the coefficient $c_{j,k}$ can only be nonzero if

$$[2^{-j}k, 2^{-j}(k+1)] \cap [0,1] \neq \emptyset.$$

We are interested in the behavior of the wavelet coefficients for large positive values of j, so we only consider functions $\psi_{j,k}$ with support in $[0,1]$; that is, for a fixed $j \in \mathbb{Z}$ we assume that $2^{-j}(k+1) \leq 1$ and that $2^{-j}k \geq 0$, i.e., that $0 \leq k \leq 2^j - 1$. Let $x_{j,k}$ be the midpoint of the interval $[2^{-j}k, 2^{-j}(k+1)]$, i.e., $x_{j,k} = 2^{-j}(k+1/2)$.

When we consider a fixed $j \in \mathbb{Z}$, there exists exactly one value of $k \in \{0, 1, \dots, 2^j - 1\}$ for which $x_0 \in [2^{-j}k, 2^{-j}(k+1)[$. For larger values of j, these intervals surrounding x_0 are getting smaller, i.e., more localized around x_0. Now, for large values of j, Walnut's argument shows that

- if k is chosen such that $x_0 \in [2^{-j}k, 2^{-j}(k+1)]$, then

$$c_{j,k} \approx \frac{1}{4}|f(x_0^-) - f(x_0^+)|\ 2^{-j/2};$$

- for all other values of k, i.e., if $x_0 \notin [2^{-j}k, 2^{-j}(k+1)]$, then

$$c_{j,k} \approx \frac{1}{4}|f'(x_{j,k})|\ 2^{-5j/2}.$$

That is, going to higher scales, i.e., increasing j, the decay of $c_{j,k}$ is much faster when we follow intervals where $x_0 \notin I_{j,k}$ than when we enter intervals for which $x_0 \in I_{j,k}$; large coefficients that persist for all scales suggest the presence of a discontinuity in the corresponding intervals. □

Walnut's argument can be extended to general wavelets with compact support:

Example 5.6.2 Consider a wavelet ψ with support on the interval $[0, M]$ for some $M \in \mathbb{N}$. Let $x_0 \in]0, M[$ and assume that $f : [0, M] \to \mathbb{C}$ is a function which is twice differentiable on the intervals $[0, x_0[$ and $]x_0, M]$; we also assume that the limits

$$f(x_0^+) := \lim_{x \to x_0^+} f(x) \text{ and } f(x_0^-) := \lim_{x \to x_0^-} f(x)$$

exist. In the following we think about f as a function defined on $\mathbb{R}$, simply by defining $f(x) = 0$ for $x \notin [0, M]$. The support of $\psi_{j,k}$ is the interval $I_{j,k} = [2^{-j}k, 2^{-j}(k+M)]$. Therefore the coefficient $c_{j,k}$ can only be nonzero if

$$[2^{-j}k, 2^{-j}(k+1)] \cap [0, M] \neq \emptyset.$$

We will only consider functions $\psi_{j,k}$ with support in $[0, M]$; that is, we consider a fixed $j \in \mathbb{Z}$ and assume that $0 \leq k \leq 2^j M - 1$. Let $x_{j,k}$ be the midpoint of the interval $[2^{-j}k, 2^{-j}(k+M)]$, i.e., $x_{j,k} = 2^{-j}(k + M/2)$.
Case 1: Assume that k is chosen such that $x_0 \notin I_{j,k}$. Then, small modifications of Walnut's proof show that

$$c_{j,k} = 2^{-3j/2} \int_{-\infty}^{\infty} x\psi(x)dx + r_{j,k}, \tag{5.26}$$

where

$$|r_{j,k}| \leq \frac{1}{24} \max_{x \in I_{j,k}} |f''(x)| \max_{x \in \mathbb{R}} |\psi(x)| \; 2^{-5j/2}. \tag{5.27}$$

Comparing the two terms in (5.26), we see that the rate of decay depends on the properties of ψ: if ψ has a vanishing moment of order 1, i.e., if

$$\int_{-\infty}^{\infty} x\psi(x)dx = 0, \tag{5.28}$$

then the first term in (5.26) (which was dominating in the case of the Haar wavelet) vanishes, and

$$|c_{j,k}| = |r_{j,k}| \leq \frac{1}{24} \max_{x \in I_{j,k}} |f''(x)| \max_{x \in \mathbb{R}} |\psi(x)| \; 2^{-5j/2}. \tag{5.29}$$

On the other hand, if the integral in (5.28) is nonzero, then the first term in (5.26) dominates, and for large j,

$$c_{j,k} \approx 2^{-3j/2} \int_{-\infty}^{\infty} x\psi(x)dx. \tag{5.30}$$

Case 2: Assume that k is chosen such that $x_0 \in I_{j,k}$. Then, again small modifications of Walnut's proof show that the first term in (5.26) dominates, and for large j,

$$c_{j,k} = (f(x_0^-) - f(x_0^+))\ 2^{-j/2} \int_0^{2^j x_0 - k} \psi(x)dx.$$

The upper limit $2^j x_0 - k$ in the integral belongs to the interval $[0, M]$; assuming it is the midpoint leads to

$$c_{j,k} \approx |f(x_0^-) - f(x_0^+)|\ 2^{-j/2} \int_0^{M/2} \psi(x)dx. \tag{5.31}$$

Comparing (5.29), (5.30) and (5.31), we again see that the coefficients decay slower if we zoom in on a point where the analyzed function has a jump, than if we zoom in on a point where the function is smooth. □

Wavelet-inspired methods can also be used to detect points where a given function is nondifferentiable. A detailed discussion can be found in the books [20], [21] by Jaffard and Meyer. Let us state a lemma, which appears in [21] with reference to an earlier paper by Freud; it tells us how a function f can be analyzed in terms of a "wavelet" ψ and the scaled versions $A^{2j}\psi(A^j x)$:

Lemma 5.6.3 *Let $A > 1$ be a constant, and let $\psi : \mathbb{R} \to \mathbb{C}$ be an integrable function for which*

$$\int_{-\infty}^{\infty} |x\psi(x)|dx < \infty \quad \textit{and} \quad \int_{-\infty}^{\infty} \psi(x)dx = \int_{-\infty}^{\infty} x\psi(x)dx = 0.$$

Let $f : \mathbb{R} \to \mathbb{C}$ be a continuous and bounded function. If f is differentiable at the point x_0, then

$$A^{2j} \int_{-\infty}^{\infty} f(x_0 - x)\psi(A^j x)dx \to 0 \textit{ as } j \to \infty. \tag{5.32}$$

If we want to analyze a function f via Lemma 5.6.3, we have to choose a function ψ which satisfies the conditions in the lemma, and for which we can calculate (or at least estimate) the expression in (5.32); if it turns out that this expression does not tend to zero as $j \to \infty$, we can conclude that f is not differentiable in x_0. This approach works very well for Weierstrass' function discussed in Example 2.5.4:

Example 5.6.4 Given constants $A > 1, B \in]0, 1[$ for which $AB \geq 1$, let

$$f(x) = \sum_{n=0}^{\infty} B^n \cos(A^n x), \ x \in \mathbb{R}.$$

Example 2.5.4 shows that f is continuous; it is also bounded, because

$$|f(x)| \leq \sum_{n=0}^{\infty} |B^n \cos(A^n x)| \leq \sum_{n=0}^{\infty} B^n = \frac{1}{1-B}.$$

Now, a clever choice of the "analyzing wavelet" ψ satisfying the conditions in Lemma 5.6.3 is made in [20]: for any $x_0 \in \mathbb{R}$ it leads to

$$A^{2j} \int_{-\infty}^{\infty} f(x_0 - x)\psi(A^j x)dx = \frac{1}{2}(AB)^j e^{iA^j x_0},$$

which does not tend to zero as $j \to \infty$. By Lemma 5.6.3, this shows that f is not differentiable at any point x_0. □

Note that we can only use Lemma 5.6.3 to obtain a negative conclusion: it can never be used to prove that a function is differentiable.

5.7 Best N-term approximation

As already described in Section 3.8, the idea behind best N-term approximation is to describe a given signal as efficiently as possible: this usually means that we want to approximate our signal via finite series expansions with as few coefficients as possible.

We shall not go into the general theory here: rather, we will look at a concrete example and examine how linear and nonlinear approximation via Fourier series and wavelet series work in that case.

Example 5.7.1 Let $x_0 \in [0,1] \setminus \mathbb{Q}$ and consider the function

$$f = \chi_{[0,x_0]}.$$

If we want to examine this function using Fourier series, we can for example ignore the values outside the interval $[-\pi, \pi]$ and perform a Fourier analysis. Considering the Fourier series in complex form, the Fourier coefficients are

$$c_n = \frac{1}{2\pi} \int_{-\pi}^{\pi} f(x)e^{-inx}dx = \frac{1}{2\pi} \int_0^{x_0} e^{inx}dx = \frac{1}{2\pi in}(e^{inx_0} - 1), \; n \in \mathbb{Z} \setminus \{0\}.$$

The absolute value of $e^{inx_0} - 1$ depends on n and can take arbitrary values in $[0, 2]$; however, it is a fair approximation to consider

$$|c_n| \sim \frac{1}{n}, n \neq 0,$$

i.e., that $|c_n|$ approximately has the size of a constant times $1/n$. Doing so, we see that if we approximate the function f by the partial sum of its

Fourier series where n ranges from $-N$ to N for some $N \in \mathbb{N}$, then

$$\begin{aligned}\left\|f - \sum_{|n|\le N} c_n e^{inx}\right\|^2 &= \left\|\sum_{|n|>N} c_n e^{inx}\right\|^2\\ &= \sum_{|n|>N} |c_n|^2\\ &\sim \sum_{N+1}^{\infty} \frac{1}{n^2}\\ &\sim \frac{1}{N}.\end{aligned}$$

A reformulation of this result says that approximation with N Fourier coefficients leads to a square-error of the size $1/N$ (times a constant).

Let us now consider approximation via the Haar wavelet. One can prove that the following argument, based on (5.23), is correct, despite the fact that f is discontinuous. Looking at the representation of f in (5.23), the natural thing to do is to order the infinite sum by starting with the small values of j. If we only have capacity to calculate or store $N = 2^J$ coefficients for some $J \in \mathbb{N}$, this means that we can consider $j = 0, 1, \ldots, J-1$; this will use

$$\sum_{j=0}^{J-1} \sum_{n=0}^{2^j-1} 1 = 1 + 2 + \cdots + 2^{J-1} = 2^J - 1 = N - 1$$

coefficients. For each of these values of j, only one of the coefficients in $\sum_{n=0}^{2^j-1} \langle f, \psi_{j,n}\rangle \psi_{j,n}(x)$ is nonzero, namely, the one corresponding to the value of n for which $x_0 \in I_{j,n}$. According to Example 5.6.1, this particular coefficient has the size $\langle f, \psi_{j,n}\rangle \sim 2^{-j/2}$; thus, approximating with N coefficients leads to an error of the size

$$\begin{aligned}\left\|f - \langle f, \chi_{[0,1[}\rangle \chi_{[0,1[} - \sum_{j=0}^{J-1}\sum_{n=0}^{2^j-1} \langle f, \psi_{j,n}\rangle \psi_{j,n}\right\|^2 &= \left\|\sum_{j=J}^{\infty}\sum_{n=0}^{2^j-1} \langle f, \psi_{j,n}\rangle \psi_{j,n}\right\|^2\\ &= \sum_{j=J}^{\infty}\sum_{n=0}^{2^j-1} |\langle f, \psi_{j,n}\rangle|^2\\ &\sim \sum_{j=J}^{\infty} 2^{-j}\\ &\sim 2^{-J}\\ &\sim \frac{1}{N}.\end{aligned}$$

The conclusion is that wavelet approximation leads to an approximation error of the same size as Fourier analysis. The picture changes completely if we consider nonlinear approximation via wavelets. As already noticed, for each level j, exactly one coefficient is nonzero: if we have capacity to calculate $N = 2^J$ coefficients, we can simply calculate the nonzero coefficients associated to the first N levels. This leads to an approximation error of the size

$$\left\| f - \langle f, \chi_{[0,1[} \rangle \chi_{[0,1[} - \sum_{j=0}^{N} \sum_{n=0}^{2^j-1} \langle f, \psi_{j,n} \rangle \psi_{j,n} \right\|^2$$

$$= \left\| \sum_{j=N+1}^{\infty} \sum_{n=0}^{2^j-1} \langle f, \psi_{j,n} \rangle \psi_{j,n} \right\|^2$$

$$= \sum_{j=N+1}^{\infty} \sum_{n=0}^{2^j-1} |\langle f, \psi_{j,n} \rangle|^2$$

$$\sim \sum_{j=N+1}^{\infty} 2^{-j}$$

$$\sim 2^{-N}.$$

That is, nonlinear approximation leads to exponential decay of the error, which is a vast improvement compared to the standard wavelet method. □

5.8 Frames

The purpose of this section is to discuss a generalization of the wavelet theory presented so far; let us first explain why we want to do so.

So far, we have been discussing wavelets ψ satisfying the two conditions (4.3) and (4.4), or, (5.2) and (5.3). However, just a glance at these conditions reveals that they can only be satisfied for very special choices of ψ. For example, (5.3) means that

$$\int_{-\infty}^{\infty} 2^{j/2+j'/2} \psi(2^j x - k) \overline{\psi(2^{j'} x - k')} dx = \begin{cases} 1 \text{ if } j = j', k = k', \\ 0 \text{ otherwise.} \end{cases}$$

It certainly needs some work to find ψ such that just this condition is satisfied. While wavelet analysis actually tells us how we can satisfy the two sets of conditions, it is also clear that it might be difficult (or even impossible) to construct wavelets having additional desirable features. A case where such a limitation appears was pointed out already by Daubechies in her book [8]:

Theorem 5.8.1 *Let $\psi \in L^2(\mathbb{R})$. Assume that ψ decays exponentially and that $\{\psi_{j,k}\}_{j,k\in\mathbb{Z}}$ is an orthonormal basis. Then ψ can not be infinitely often differentiable with bounded derivatives.*

Another limitation appears if we want ψ to be symmetric around a certain axis; see again [8] for details:

Theorem 5.8.2 *There exists only one wavelet ψ (generated by an MRA with a real scaling function) which has compact support and a symmetry or antisymmetry axis, namely, the Haar wavelet.*

These are just a few examples of no-go results for wavelets. A way to gain extra flexibility is to weaken the conditions we put on a wavelet. In order to do so, we need to discuss why we introduced the conditions used so far.

The key point in this book has been that series expansions are useful because they deliver convenient expressions for the expanded function, often with good approximation properties. From this perspective, it is natural to discuss orthonormal bases, because they always lead to such expansions in the relevant function spaces. Formulated differently: we are not so much interested in the orthonormal basis property in itself as in the fact that it leads to a nice expansion.

As a natural continuation of this path of thinking, we consider the following program for more general expansions via wavelet systems:

- We insist on each $f \in L^2(\mathbb{R})$ having a series expansion of the type

$$f(x) = \sum_{j,k\in\mathbb{Z}} c_{j,k}\psi_{j,k}(x) \tag{5.33}$$

  for certain coefficients $\{c_{j,k}\}_{j,k\in\mathbb{Z}}$;

- we no longer insist on $\{\psi_{j,k}\}_{j,k\in\mathbb{Z}}$ being an orthonormal basis for $L^2(\mathbb{R})$.

Since this setup is more general than the orthonormal basis constructions considered so far, the coefficients satisfying (5.33) might not be given by the expression in (4.5).

There are in fact several ways of obtaining such more general expansions. One of them is to ask for $\{\psi_{j,k}\}_{j,k\in\mathbb{Z}}$ being a *Riesz basis.* We will not give the general definition, but just mention that this guarantees a series expansion of the type (4.3) for each $f \in L^2(\mathbb{R})$. Furthermore, a construction by Cohen, Daubechies, and Feauveau [5] leads to coefficients $c_{j,k}$ having a nice expression: in fact, for a certain function $\tilde{\psi}$ they have the form

$$c_{j,k} = \int_{-\infty}^{\infty} f(x)\overline{\tilde{\psi}_{j,k}(x)}dx.$$

To separate these wavelets from the one introduced so far, they are called *biorthogonal wavelets*; as discussed in Section 5.8 they have already found important applications.

A way to obtain even more flexibility is to consider *frames* (a Riesz basis is a special case of a frame). For later use we define frames in general Hilbert spaces.

Definition 5.8.3 *Let $\{f_k\}_{k=1}^{\infty}$ be a sequence in a Hilbert space $\mathcal{H}$. Then we say that*

(i) $\{f_k\}_{k=1}^{\infty}$ is a frame for $\mathcal{H}$ if there exist constants $A, B > 0$ such that

$$A\,||f||^2 \leq \sum_{k=1}^{\infty} |\langle f, f_k\rangle|^2 \leq B\,||f||^2, \;\forall f \in \mathcal{H}; \tag{5.34}$$

(ii) $\{f_k\}_{k=1}^{\infty}$ is a tight frame if (5.34) holds with $A = B$.

Before we proceed with a technical discussion, we mention that frames have an interesting history. They were introduced in full generality in the paper [12] by Duffin and Schaeffer, dating back to 1952. However, apparently nobody realized the potential of frames at that time; at least, frames disappeared from the literature for almost 30 years. The next treatment came in 1980, where Young described frames and their properties in his book [28]. Then, in the middle of the eighties as the wavelet theory started to develop, Grossmann observed that frames would be useful in order to obtain expansions in terms of the functions in a wavelet system, see [9]; since then, frame theory has developed rapidly. Today, frame theory is a well established research area in harmonic analysis; see, e.g., the book [3].

One can prove that if $\{f_k\}_{k=1}^{\infty}$ is a frame, then there exists a sequence $\{g_k\}_{k=1}^{\infty}$ in $\mathcal{H}$ such that

$$f = \sum_{k=1}^{\infty} \langle f, g_k\rangle f_k, \;\forall f \in \mathcal{H}. \tag{5.35}$$

Thus, for a frame we always obtain an expansion which is similar to the expansion (3.17) via an orthonormal basis: the difference is that the coefficients in (5.35) have to be calculated via the sequence $\{g_k\}_{k=1}^{\infty}$, while the coefficients in (3.17) are given via inner products between the signal f and the elements in the orthonormal basis. A sequence $\{g_k\}_{k=1}^{\infty}$ satisfying (5.35) is called a *dual* of $\{f_k\}_{k=1}^{\infty}$.

The condition (5.2) shows that the class of frames contains all orthonormal bases. That is, going from orthonormal bases to frames giyes us more flexibility: we might be able to make frame constructions with certain properties which are impossible for orthonormal bases.

An important feature of a frame $\{f_k\}_{k=1}^{\infty}$ is that several choices of $\{g_k\}_{k=1}^{\infty}$ are possible, except in the special case where $\{f_k\}_{k=1}^{\infty}$ is a basis. This makes it possible to search for the dual $\{g_k\}_{k=1}^{\infty}$ which is most convenient in a given situation. There are in fact cases in wavelet theory where this feature is very convenient:

Example 5.8.4 Assume that a wavelet system $\{\psi_{j,k}\}_{j,k\in\mathbb{Z}}$ forms a frame for $L^2(\mathbb{R})$. Then the general frame theory tells us that there exists a dual, i.e., a double-indiced family $\{g^{j,k}\}_{j,k\in\mathbb{Z}}$ of functions in $L^2(\mathbb{R})$ such that

$$f = \sum_{j,k\in\mathbb{Z}} \langle f, g^{j,k}\rangle \psi_{j,k}, \ \forall f \in L^2(\mathbb{R}).$$

However, in contrast to the orthonormal case, the family $\{g^{j,k}\}_{j,k\in\mathbb{Z}}$ might not be a wavelet system! That is, nothing guarantees that $\{g^{j,k}\}_{j,k\in\mathbb{Z}}$ is generated by scalings and translations of a single function. This is very inconvenient, e.g., in case we have to store information about $\{g^{j,k}\}_{j,k\in\mathbb{Z}}$ electronically (see page 88).

However, in certain cases we can *find* a dual which has wavelet structure. The research paper [10] contains a concrete example of a wavelet frame, where certain duals have wavelet structure and certain duals do not. □

For general frames, it is frequently difficult to find a suitable family $\{g_k\}_{k=1}^{\infty}$ explicitly, and therefore it might be difficult to find the coefficients $\langle f, g_k\rangle$ in (5.35). This is the reason for the interest in tight frames: if $\{f_k\}_{k=1}^{\infty}$ is a tight frame, one can prove that we can take $g_k = \frac{1}{A} f_k$, which immediately leads to the series expansion

$$f = \frac{1}{A}\sum_{k=1}^{\infty} \langle f, f_k\rangle f_k, \ \forall f \in \mathcal{H}.$$

In this expression, no extra effort (compared to the use of an orthonormal basis) is needed in order to calculate the coefficients.

Frames actually lead to more flexibility, compared to use of orthonormal bases. Already on page 92 we mentioned that the wavelet system generated by the Gaussian $\psi(x) = e^{-x^2}$ does not form an orthonormal basis for $L^2(\mathbb{R})$. However, it turns out that $\{\psi_{j,k}\}_{j,k\in\mathbb{Z}}$ is a frame. Note that this wavelet system has all the features described in Theorem 5.8.1, except that it forms a frame rather than an orthonormal basis; thus, in cases where we only need the expansion property, we can use this wavelet system to avoid the obstacle in Theorem 5.8.1.

As a follow-up to a series of papers by Ron and Shen, Chui et al. [6] as well as Daubechies et al. [11] have recently constructed tight frames of the form $\{\psi^1_{j,k}\}_{j,k\in\mathbb{Z}} \cup \{\psi^2_{j,k}\}_{j,k\in\mathbb{Z}}$; that is, systems obtaining by combining the wavelet systems associated with *two* functions ψ^1, ψ^2. Figure 5.9.1 shows an example of how such functions can be chosen. They are splines (in fact, piecewise polynomials of degree at most 1) and easy and explicit expressions exist. Comparing with Daubechies' wavelets in Figure 4.1.9

we immediately see that we are dealing with much simpler functions. In fact, the constructions in [6] and [11] can be used to construct tight frames, where the generators ψ^1, ψ^2 are splines of any desired order, having compact support. A further advantage is that there exist concrete and convenient expressions for these splines. In contrast, Daubechies' wavelets are not given explicitly via elementary functions: one has to find their pointwise values via a recursion algorithm.

We also note that the shortcoming for wavelets discussed in Theorem 5.8.2 does not appear for the wavelet frame constructions in [6] and [11]: the functions ψ^1, ψ^2 in Figure 5.9.1 have compact support and a symmetry axis, respectively an antisymmetry axis.

5.9 Gabor systems

In this section we give a short introduction to Gabor systems and a few of their mathematical properties. Gabor systems arise naturally if we want to use Fourier analysis to construct an orthonormal basis for $L^2(\mathbb{R})$; we will use this approach, and give the formal definition later.

In Section 3.5 we have seen that the functions $\{\frac{1}{\sqrt{2\pi}}e^{imx}\}_{m\in\mathbb{Z}}$ form an orthonormal basis for $L^2(-\pi,\pi)$; since they are periodic with period 2π, they actually form an orthonormal basis for $L^2(-\pi+2\pi n, \pi+2\pi n)$ for any $n \in \mathbb{Z}$. If we want to put emphasis on the fact that we look at the exponential functions on the interval $[-\pi+2\pi n, \pi+2\pi n[$, we can also write that

$$\left\{\frac{1}{\sqrt{2\pi}}e^{imx}\chi_{[-\pi+2\pi n,\pi+2\pi n[}(x)\right\}_{m\in\mathbb{Z}}$$

is an orthonormal basis for $L^2(-\pi+2\pi n, \pi+2\pi n)$. Now observe that the intervals $[-\pi+2\pi n, \pi+2\pi n[$, $n \in \mathbb{Z}$, form a partition of $\mathbb{R}$: they are disjoint and cover the entire axis $\mathbb{R}$. This implies that the union of these bases, i.e., the family

$$\left\{\frac{1}{\sqrt{2\pi}}e^{imx}\chi_{[-\pi+2\pi n,\pi+2\pi n[}(x)\right\}_{m,n\in\mathbb{Z}},$$

is an orthonormal basis for $L^2(\mathbb{R})$. We can write this orthonormal basis on a slightly more convenient form as

$$\left\{\frac{1}{\sqrt{2\pi}}e^{imx}\chi_{[-\pi,\pi[}(x-2\pi n)\right\}_{m,n\in\mathbb{Z}}. \tag{5.36}$$

The system in (5.36) is the simplest case of a Gabor system. Basically, (5.36) consists of the function $\frac{1}{\sqrt{2\pi}}\chi_{[-\pi+2\pi n,\pi+2\pi n[}(x)$ and translated versions and *modulated versions*, i.e., functions which are multiplied by complex exponential functions. A general Gabor system is obtained by replacing

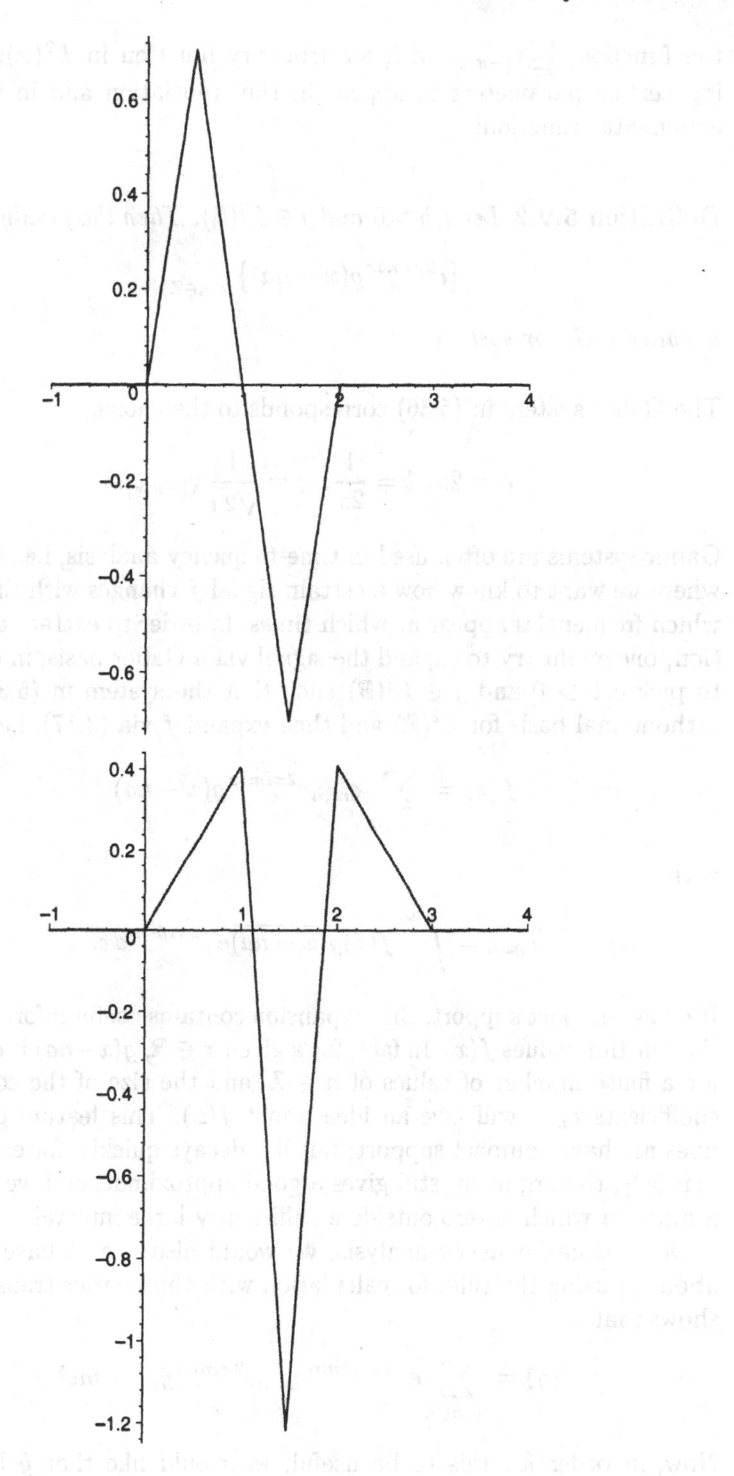

Figure 5.9.1 *Two functions ψ^1, ψ^2 for which $\{\psi^1_{j,k}\}_{j,k\in\mathbb{Z}} \cup \{\psi^2_{j,k}\}_{j,k\in\mathbb{Z}}$ forms a frame for $L^2(\mathbb{R})$.*

the function $\frac{1}{\sqrt{2\pi}}\chi_{[-\pi,\pi[}$ with an arbitrary function in $L^2(\mathbb{R})$, and allowing certain parameters to appear in the translation and in the complex exponential function:

Definition 5.9.2 *Let $a, b > 0$ and $g \in L^2(\mathbb{R})$. Then the family of functions*

$$\left\{e^{2\pi imbx}g(x - na)\right\}_{m,n\in\mathbb{Z}} \tag{5.37}$$

is called a Gabor system.

The Gabor system in (5.36) corresponds to the choice

$$a = 2\pi,\ b = \frac{1}{2\pi},\ g = \frac{1}{\sqrt{2\pi}}\chi_{[-\pi,\pi[}.$$

Gabor systems are often used in time-frequency analysis, i.e., in situations where we want to know how a certain signal f changes with time, and also which frequencies appear at which times. In order to extract this information, one might try to expand the signal via a Gabor basis; in other words, to pick $a, b > 0$ and $g \in L^2(\mathbb{R})$ such that the system in (5.37) forms an orthonormal basis for $L^2(\mathbb{R})$ and then expand f via (3.17), i.e., write

$$f(x) = \sum_{m,n\in\mathbb{Z}} c_{m,n}e^{2\pi imbx}g(x - na) \tag{5.38}$$

with

$$c_{m,n} = \int_{-\infty}^{\infty} f(x)\overline{g(x - na)}e^{-2\pi imbx}dx.$$

If g has compact support, this expansion contains useful information about the function values $f(x)$. In fact, for a given $x \in \mathbb{R}$, $g(x-na)$ is only nonzero for a finite number of values of $n \in \mathbb{Z}$, and the size of the corresponding coefficients $c_{m,n}$ will give an idea about $f(x)$. This feature gets lost if g does not have compact support; but if g decays quickly, for example exponentially, the argument still gives a good approximation if we replace g by a function which is zero outside a sufficiently large interval.

Doing time-frequency analysis, we would also like to have information about $\hat{f}$; using the rules for calculation with the Fourier transform indeed shows that

$$\hat{f}(\gamma) = \sum_{m,n\in\mathbb{Z}} e^{-2\pi iabmn}c_{m,n}e^{2\pi ina\gamma}\hat{g}(\gamma - mb).$$

Now, in order for this to be useful, we would like that $\hat{g}$ has compact support, or at least decays fast. However, here we touch a famous problem, formulated via the *Balian–Low theorem:*

Theorem 5.9.3 *Let $g \in L^2(\mathbb{R})$ and $a, b > 0$ be given. If the Gabor system (5.37) is an orthonormal basis for $L^2(\mathbb{R})$, then*

$$\left(\int_{-\infty}^{\infty} |xg(x)|^2 dx\right)\left(\int_{-\infty}^{\infty} |\gamma\hat{g}(\gamma)|^2 d\gamma\right) = \infty. \tag{5.39}$$

For a proof of the Balian–Low theorem we refer to [18]. In words, the Balian–Low theorem means that a function g generating a Gabor basis can not be well localized in both time and frequency, in the sense that both g and $\hat{g}$ have compact support or decay quickly. For example, it is not possible that g and $\hat{g}$ satisfy estimates like

$$|g(x)| \le \frac{C}{1+x^2}, \quad |\hat{g}(\gamma)| \le \frac{C}{1+\gamma^2}$$

simultaneously.

If fast decay of g and $\hat{g}$ is needed, we have to ask whether we need all the properties characterizing an orthonormal basis or whether we can relax some of them. The property we want to keep is that every $f \in L^2(\mathbb{R})$ has an expansion in terms of modulated and translated versions of the function g, as in (5.38). It turns out that the expansion property actually can be combined with g and $\hat{g}$ having very fast decay: what we have to do is to allow the Gabor system to be a frame rather than an orthonormal basis:

Example 5.9.4 Let $a, b > 0$ and consider $g(x) = e^{-x^2}$. Then one can prove (it is difficult and not meant to be an exercise!) that the generated Gabor system is a frame if and only if $ab < 1$. The frames we obtain for $ab < 1$ are very well localized in time (by their definition) and in frequency, because

$$\hat{g}(\gamma) = \sqrt{\pi}e^{-\pi^2 x^2}.$$

□

As we have seen several times before, practical calculations always have to be performed on finite systems of functions; for Gabor systems this means that at a certain point we will have to restrict our attention to a finite family of the form

$$\left\{e^{2\pi imbx} g(x - na)\right\}_{|m|,|n| \le N} \tag{5.40}$$

(or a subset hereof) for some $N \in \mathbb{N}$. This family spans a finite-dimensional subspace V of $L^2(\mathbb{R})$, and we can deal with it via linear algebra. An interesting and somehow surprising fact is that the vectors always form a basis for V, i.e., they are automatically linearly independent! This result was proved in some special cases by Heil et al. in [17] and conjectured to hold generally; the full proof was given later by Linnell [24].

However, as a final comment we note that the original conjecture by Heil et al. is even more general: it claims that if $\{(\mu_k, \lambda_k)\}_{k=1}^N$ is an arbitrary collection of distinct points in $\mathbb{R}^2$ and $g \neq 0$, then the *generalized Gabor system*

$$\{e^{2\pi i\lambda_k x} g(x-\mu_k)\}_{k=1}^{N}$$

is linearly independent. This conjecture is still open, despite the fact that several researchers have tried hard to prove or disprove it.

5.10 Exercises

5.1 Consider the function

$$f(x) = \sin \pi x \ \chi_{[0,1]}(x).$$

(i) Make drafts (e.g., via Maple or Mathematica) of some of the functions f_k introduced in Section 5.4 and compare with f.

(ii) Compare the approximations in (i) with the partial sums of the power series for the function $x \mapsto \sin \pi x$.

5.2 Consider the function

$$f(x) = e^x \ \chi_{[0,1]}(x).$$

(i) Make drafts (e.g., via Maple or Mathematica) of some of the functions f_k introduced in Section 5.4 and compare with f.

(ii) Compare the approximations in (i) with the partial sums of the power series for the exponential function.

(iii) Compare the approximations in (i) with the partial sums of the Fourier series for the 2π-periodic function given by

$$g(x) = e^x, \ x \in [-\pi, \pi[.$$

Appendix A

A.1 Definitions and notation

The absolute value of a real number x is defined by:

$$|x| = \begin{cases} x & \text{if } x \geq 0; \\ -x & \text{if } x < 0 \,. \end{cases}$$

For a complex number $a + ib$ where $a, b \in \mathbb{R}$,

$$|a + ib| = \sqrt{a^2 + b^2}.$$

We will often use the *triangular inequality*, which says that

$$|x + y| \leq |x| + |y| \text{ for all } x, y \in \mathbb{C}.$$

A subset E of $\mathbb{R}$ is *bounded above* if there exists $\beta \in \mathbb{R}$ such that

$$x \leq \beta, \ \forall x \in E; \tag{A.1}$$

the smallest number β satisfying (A.1) is called the *supremum of* E, and is written $\sup E$, or, $\sup_{x \in E} x$. Similarly, the set E is bounded below if there exists $\alpha \in \mathbb{R}$ such that

$$\alpha \leq x, \ \forall x \in E;$$

and the largest number with this property is called the *infimum of* E, $\inf E$. For example,

$$\sup]0,4] = 4, \quad \sup[-2,5[= 5, \quad \inf]-2,1] = -2, \quad \inf(\mathbb{Q} \cap [\pi, 7]) = \pi.$$

Let $I \subset \mathbb{R}$ be an interval. A function $f : I \to \mathbb{R}$ is *continuous* at a point $x_0 \in I$ if for any given $\epsilon > 0$ we can find $\delta > 0$ such that

$$|f(x) - f(x_0)| \leq \epsilon \text{ for all } x \in I \text{ for which } |x - x_0| \leq \delta. \tag{A.2}$$

We say that f is continuous if f is continuous at every point in I.

A continuous function on a bounded and closed interval $[a, b]$ is *uniformly continuous*; that is, for each $\epsilon > 0$ we can find $\delta > 0$ such that

$$|f(x) - f(x_0)| \leq \epsilon \text{ for all } x, x_0 \in [a, b] \text{ for which } |x - x_0| \leq \delta. \tag{A.3}$$

Observe the difference between (A.2) and (A.3): in the first case x_0 is a *fixed* point in the interval $[a, b]$, while in the second case we also allow x_0 to vary. Therefore (A.3) is in general more restrictive than (A.2).

A continuous real-valued function defined on a bounded and closed interval $[a, b]$ assumes its supremum, i.e., there exists $x_0 \in [a, b]$ such that

$$f(x_0) = \sup \{f(x) \mid x \in [a, b]\}.$$

Similarly, the function assumes its infimum. If $x_0 \in]a, b[$ is a point in which a differentiable function $f : [a, b] \to \mathbb{R}$ assumes its maximal or minimal value, then $f'(x_0) = 0$.

A.2 Proof of Weierstrass' theorem

For sake of simplicity we assume that $I = [0, 1]$; the general case can be obtained from this case via a scaling. We define a sequence of polynomials $Q_N, N \in \mathbb{N}$, by

$$Q_N(x) = \sum_{n=0}^{N} f(\frac{n}{N}) \binom{N}{n} x^n (1-x)^{N-n}, \tag{A.4}$$

where

$$\binom{N}{n} = \frac{N!}{n!(N-n)!}.$$

In order to estimate the quantity $|f(x) - Q_N(x)|$ we first rewrite $f(x)$ with help of the binomial formula. The trick is to multiply $f(x)$ by the number 1, which we can write as

$$\begin{aligned} 1 &= (x + (1-x))^N \\ &= \sum_{n=0}^{N} \binom{N}{n} x^n (1-x)^{N-n}. \end{aligned}$$

Hereby we get

$$\begin{aligned} f(x) &= f(x)\,(x+(1-x))^N \\ &= f(x)\sum_{n=0}^{N}\binom{N}{n}x^n(1-x)^{N-n} \\ &= \sum_{n=0}^{N} f(x)\binom{N}{n}x^n(1-x)^{N-n}. \end{aligned}$$

Thus,

$$\begin{aligned} & f(x)-Q_N(x) \\ = & \sum_{n=0}^{N} f(x)\binom{N}{n}x^n(1-x)^{N-n} - \sum_{n=0}^{N} f(\tfrac{n}{N})\binom{N}{n}x^n(1-x)^{N-n} \\ = & \sum_{n=0}^{N}(f(x)-f(\tfrac{n}{N}))\binom{N}{n}x^n(1-x)^{N-n}. \end{aligned} \tag{A.5}$$

For a given $\epsilon > 0$, we will show that the degree of polynomial Q_N can be chosen such that $|f(x) - Q_N(x)| \le \epsilon$ for all $x \in [0,1]$. We will use the fact that the function f is uniformly continuous on the interval $[0,1]$; by (A.3) this means that we can find a $\delta > 0$ such that

$$x,y \in [0,1], |x-y| < \delta \Rightarrow |f(x)-f(y)| < \frac{\epsilon}{2}. \tag{A.6}$$

Fix $x \in [0,1]$. We first divide the index set for the sum (A.5) into two sets:

$$f(x)-Q_N(x) = \sum_{n=0}^{N}(f(x)-f(\tfrac{n}{N}))\binom{N}{n}x^n(1-x)^{N-n} \tag{A.7}$$

$$= \sum_{|x-\frac{n}{N}|<\delta}(f(x)-f(\tfrac{n}{N}))\binom{N}{n}x^n(1-x)^{N-n} \tag{A.8}$$

$$+ \sum_{|x-\frac{n}{N}|\ge\delta}(f(x)-f(\tfrac{n}{N}))\binom{N}{n}x^n(1-x)^{N-n}. \tag{A.9}$$

In the following we investigate (A.8) and (A.9) separately. We begin by considering (A.8). By the triangle inequality

$$\begin{aligned} & \left|\sum_{|x-\frac{n}{N}|<\delta}(f(x)-f(\tfrac{n}{N}))\binom{N}{n}x^n(1-x)^{N-n}\right| \\ \le & \sum_{|x-\frac{n}{N}|<\delta}\left|f(x)-f(\tfrac{n}{N})\right|\binom{N}{n}x^n(1-x)^{N-n}. \end{aligned}$$

It follows by our choice of δ in (A.6) that

$$\begin{aligned}
&\sum_{|x-\frac{n}{N}|<\delta} \left|f(x) - f(\frac{n}{N})\right| \binom{N}{n} x^n(1-x)^{N-n} \\
&\leq \frac{\epsilon}{2} \sum_{|x-\frac{n}{N}|<\delta} \binom{N}{n} x^n(1-x)^{N-n} \\
&\leq \frac{\epsilon}{2} \sum_{n=0}^{N} \binom{N}{n} x^n(1-x)^{N-n} \\
&= \frac{\epsilon}{2}(x+(1-x))^N \\
&= \frac{\epsilon}{2}.
\end{aligned}$$

Therefore we obtain the following estimate for (A.8):

$$\left| \sum_{|x-\frac{n}{N}|<\delta} f(x) - f(\frac{n}{N}) \binom{N}{n} x^n(1-x)^{N-n} \right| \leq \frac{\epsilon}{2}. \tag{A.10}$$

Next, we consider (A.9). First we notice that

$$\begin{aligned}
&\left| \sum_{|x-\frac{n}{N}|\geq\delta} f(x) - f(\frac{n}{N}) \binom{N}{n} x^n(1-x)^{N-n} \right| \\
&\leq \sum_{\frac{(x-\frac{n}{N})^2}{\delta^2}\geq 1} \left|f(x) - f(\frac{n}{N})\right| \binom{N}{n} x^n(1-x)^{N-n}.
\end{aligned}$$

Let $M = \max_{x\in[0,1]} |f(x)|$. By the triangle inequality,

$$|f(x) - f(\frac{n}{N})| \leq |f(x)| + |f(\frac{n}{N})| \leq 2M. \tag{A.11}$$

Using (A.11) it follows that

$$\begin{aligned}
&\sum_{\frac{(x-\frac{n}{N})^2}{\delta^2}\geq 1} \left|f(x) - f(\frac{n}{N})\right| \binom{N}{n} x^n(1-x)^{N-n} \\
&\leq 2M \sum_{\frac{(x-\frac{n}{N})^2}{\delta^2}\geq 1} \binom{N}{n} x^n(1-x)^{N-n} \\
&\leq 2M \sum_{\frac{(x-\frac{n}{N})^2}{\delta^2}\geq 1} \frac{(x-\frac{n}{N})^2}{\delta^2} \binom{N}{n} x^n(1-x)^{N-n} = (*);
\end{aligned}$$

in the last inequality we used that $\frac{(x-\frac{n}{N})^2}{\delta^2} \geq 1$ for all the values of n which appear in the summation. Furthermore, the sum increases if we sum over all $n = 0, 1, \ldots, N$ rather than just the values of n for which $\frac{(x-\frac{n}{N})^2}{\delta^2} \geq 1$; thus

$$
\begin{aligned}
(*) &\leq 2M \sum_{n=0}^{N} \frac{(x-\frac{n}{N})^2}{\delta^2} \binom{N}{n} x^n (1-x)^{N-n} \\
&\leq \frac{2M}{\delta^2} \sum_{n=0}^{N} \left(x - \frac{n}{N}\right)^2 \binom{N}{n} x^n (1-x)^{N-n} = (**).
\end{aligned}
$$

Note that $\sum_{n=0}^{N} (n - Nx)^2 \binom{N}{n} x^n (1-x)^{N-n}$ is the variance for the binomial distribution with parameter N and probability parameter x; thus, it equals $Nx(1-x)$. Thus,

$$\sum_{n=0}^{N} \left(x - \frac{n}{N}\right)^2 \binom{N}{n} x^n (1-x)^{N-n} = \frac{1}{N} x(1-x),$$

and

$$(**) = \frac{2M}{N\delta^2} x(1-x).$$

Since we only consider $x \in [0, 1]$,

$$\frac{2M}{N\delta^2} x(1-x) \leq \frac{M}{2N\delta^2}.$$

Altogether we now have that the term in (A.9) can be estimated by

$$\left| \sum_{|x-\frac{n}{N}| \geq \delta} f(x) - f\left(\frac{n}{N}\right) \binom{N}{n} (1-x)^{N-n} \right| \leq \frac{M}{2N\delta^2}. \quad \text{(A.12)}$$

Let us now return to the expression for $f(x) - Q_N(x)$ in (A.7). According to our estimates for the terms in (A.8) and (A.9) in (A.10) and (A.12), we have that

$$|f(x) - Q_N(x)| \leq \frac{\epsilon}{2} + \frac{M}{2N\delta^2}.$$

This holds for all values of x. Now choose $N \in \mathbb{N}$ such that $\frac{M}{2N\delta^2} < \frac{\epsilon}{2}$, i.e., $N > \frac{M}{\delta^2 \epsilon}$. Then

$$|f(x) - Q_N(x)| < \epsilon \text{ for all } x \in [0, 1]. \qquad \square$$

A.3 Proof of Taylor's theorem

Before we prove Taylors theorem we state some results which will be used in the proof. We begin with *Rolle's theorem:*

Theorem A.3.1 *Assume that the function $f : [a,b] \to \mathbb{R}$ is differentiable and that $f(a) = f(b) = 0$. Then there exists a point $\xi \in]a,b[$, for which $f'(\xi) = 0$.*

Proof: The function f has a maximal value as well as a minimal value in the interval $[a,b]$. If the maximal value equals the minimal value, then f is constant, and $f'(x) = 0$ for all $x \in]a,b[$. On the other hand, if the minimal value is different from the maximal value, then there exists $\xi \in]a,b[$ where f assumes an extremal value, because of the assumption $f(a) = f(b)$; but this implies that $f'(\xi) = 0$. □

Lemma A.3.2 *Assume that $f : [a,b] \to \mathbb{R}$ is arbitrarily often differentiable, and let x, x_0 be arbitrary points in $[a,b]$, with $x > x_0$. Assume that for some $N \in \mathbb{N}$,*

$$f(x) = 0 \text{ and } f^{(j)}(x_0) = 0 \text{ for all } j \in \{0,1,\dots,N\}. \tag{A.13}$$

Then there exists $\xi \in]x_0, x[$ such that $f^{(N+1)}(\xi) = 0$.

Proof: The proof consists of a continued application of Rolle's theorem. We first prove the result for $N = 0$, and then move on with $N = 1$, etc. Observe that when we want to prove the theorem for one value of N, the assumption (A.13) is available for all $j \in \{0,1,\dots,N\}$; this will play a key role in the proof.

For $N = 0$, our assumption is that $f(x) = f(x_0) = 0$. We shall prove the existence of a point $\xi_1 \in]x_0, x[$ for which$f'(\xi_1) = 0$; but this is exactly the conclusion in Rolle's theorem.

Now consider $N = 1$; in this case, the assumption (A.13) is available for $j = 0$ and $j = 1$. Via the assumption for $j = 0$ we just proved the existence of $\xi_1 \in]x_0, x[$, for which $f'(\xi_1) = 0$. Now, if we use the assumption for $j = 1$ we also have that $f'(x_0) = 0$. Applying Rolle's theorem again, this time to the function f' and the interval $[x_0, \xi_1]$, we conclude that there exists $\xi_2 \in]x_0, \xi_1[$ such that $f''(\xi_2) = 0$. This is exactly the conclusion we wanted to obtain for $N = 1$.

We now consider the general case, where $N > 1$. Thus, the assumption (A.13) means that

$$f(x) = f(x_0) = f'(x_0) = f''(x_0) = \cdots = f^{(N)}(x_0) = 0.$$

Applying Rolle's theorem to the function f and the interval $[x_0, x]$, we can find $\xi_1 \in]x_0, x[$ for which $f'(\xi_1) = 0$. Applying Rolle's theorem to f' and the interval $[x_0, \xi_1]$ leads to a point $\xi_2 \in]x_0, \xi_1[$ for which $f''(\xi_2) = 0$. Applying this procedure $N + 1$ times finally leads to a point ξ_{N+1} in $]x_0, \xi_N[$ for which $f^{(N+1)}(\xi_{N+1}) = 0$. □

Now we only need one step before we are ready to prove Taylor's theorem. As before, we let P_N denote the Nth Taylor polynomial for f at x_0:

Theorem A.3.3 *Assume that $f : [a,b] \to \mathbb{R}$ is arbitrarily often differentiable, and let $x_0 \in]a,b[$, $N \in \mathbb{N}$. Then, for each $x \in [a,b]$ we can find ξ between x and x_0 such that*

$$f(x) = P_N(x_0) + \frac{f^{(N+1)}(\xi)}{(N+1)!}(x-x_0)^{N+1}.$$

Proof: Fix $x \in [a,b]$. The statement certainly holds if $x = x_0$, so we assume that $x_0 < x$ (the proof is similar if $x_0 > x$). Consider the function ϕ defined via

$$\phi(t) = f(t) - P_N(t) - K(t-x_0)^{N+1}; \qquad \text{(A.14)}$$

here K is a constant, which we choose such that $\phi(x) = 0$, i.e.,

$$K = \frac{f(x) - P_N(x)}{(x-x_0)^{N+1}}.$$

Differentiating leads to

$$\phi'(t) = f'(t) - P_N'(t) - (N+1)K(t-x_0)^N,$$

and more generally, for each integer $j \in \{1,2,3,\dots,N+1\}$,

$$\begin{aligned} \phi^{(j)}(t) &= f^{(j)}(t) - P_N^{(j)}(t) \\ &\quad -(N+1)N(N-1)\cdots(N-j+2)K(t-x_0)^{N-j+1}. \end{aligned}$$

In particular,

$$\phi^{(N+1)}(t) = f^{(N+1)}(t) - P_N^{(N+1)}(t) - (N+1)!K. \qquad \text{(A.15)}$$

By definition,

$$\begin{aligned} P_N(t) &= f(x_0) + \frac{f'(x_0)}{1!}(t-x_0) + \frac{f''(x_0)}{2!}(t-x_0)^2 \\ &\quad + \frac{f^{(3)}(x_0)}{3!}(t-x_0)^3 + \cdots + \frac{f^{(N)}(x_0)}{N!}(t-x_0)^N. \end{aligned}$$

In this expression the first term is a constant, so

$$\begin{aligned} P_N'(t) &= f'(x_0) + \frac{f''(x_0)}{2!}2(t-x_0) \\ &\quad + \frac{f^{(3)}(x_0)}{3!}3(t-x_0)^2 + \cdots + \frac{f^{(N)}(x_0)}{N!}N(t-x_0)^{N-1} \\ &= f'(x_0) + f''(x_0)(t-x_0) \\ &\quad + \frac{f^{(3)}(x_0)}{2!}(t-x_0)^2 + \cdots + \frac{f^{(N)}(x_0)}{(N-1)!}(t-x_0)^{N-1}. \end{aligned}$$

Differentiating once more,

$$\begin{aligned} P_N''(t) &= f''(x_0) + \frac{2}{2!}f^{(3)}(x_0)(t-x_0) \\ &\quad + \cdots + \frac{(N-1)}{(N-1)!}f^{(N)}(x_0)(t-x_0)^{N-2} \\ &= f''(x_0) + f^{(3)}(x_0)(t-x_0) + \cdots + \frac{f^{(N)}(x_0)}{(N-2)!}(t-x_0)^{N-2}. \end{aligned}$$

More generally, for $j \in \{1, 2, 3, \ldots, N\}$,

$$\begin{aligned} P_N^{(j)}(t) &= f^{(j)}(x_0) + f^{(j+1)}(x_0)(t-x_0) \\ &+ \frac{f^{(j+2)}(x_0)}{2!}(t-x_0)^2 + \cdots + \frac{f^{(N)}(x_0)}{(N-j)!}(t-x_0)^{N-j}. \end{aligned}$$

Note that

$$P_N^{(N+1)}(t) = 0, \ \forall t \in \mathbb{R}.$$

For $j \in \{0, 1, \ldots, N\}$, we have $P_n^{(j)}(x_0) = f^{(j)}(x_0)$, and therefore $\phi^{(j)}(x_0) = 0$. This means that the function ϕ satisfies the conditions in Lemma A.3.2. Thus, there exists $\xi \in]x_0, x[$ for which $\phi^{(N+1)}(\xi) = 0$. But according to (A.15),

$$\phi^{(N+1)}(\xi) = f^{(N+1)}(\xi) - (N+1)!K,$$

so

$$f^{(N+1)}(\xi) = (N+1)!K \text{ and } K = \frac{f^{(N+1)}(\xi)}{(N+1)!}.$$

Using the expression (A.14) for ϕ and that $\phi(x) = 0$, we see that

$$f(x) = P_N(x) + \frac{f^{(N+1)}(\xi)}{(N+1)!}(x-x_0)^{N+1}.$$

□

Here is finally the proof for Taylor's theorem, in the case $I = [a, b]$:

Proof of Theorem 1.3.6: Let $N \in \mathbb{N}$ and $x \in]a, b[$ be given. According to Theorem A.3.3 we can find $\xi \in]a, b[$ such that

$$f(x) = \sum_{n=0}^{N} \frac{f^{(n)}(x_0)}{n!}(x-x_0)^n + \frac{f^{(N+1)}(\xi)}{(N+1)!}(x-x_0)^{N+1}.$$

Thus,

$$f(x) - \sum_{n=0}^{N} \frac{f^{(n)}(x_0)}{n!}(x-x_0)^n = \frac{f^{(N+1)}(\xi)}{(N+1)!}(x-x_0)^{N+1}.$$

By assumption we have that $|f^{(N+1)}(x)| \leq C$ for all $x \in]a,b[$, so this implies that

$$\begin{aligned}\left|f(x) - \sum_{n=0}^{N} \frac{f^{(n)}(x_0)}{n!}(x-x_0)^n\right| &= \left|\frac{f^{(N+1)}(\xi)}{(N+1)!}(x-x_0)^{N+1}\right| \\ &\leq \frac{C}{(N+1)!}|x-x_0|^{N+1}.\end{aligned}$$

□

A.4 Infinite series

Proof of Theorem 2.1.9 First we assume that (2.5) is satisfied with $C < 1$. Then we can find $d \in]0,1[$ such that $\left|\frac{a_{n+1}}{a_n}\right| \leq d$ for all sufficiently large values of n, say, for all $n \geq N$. Now, considering $n > N$,

$$\begin{aligned}|a_n| &= \left|\frac{a_n}{a_{n-1}} \cdot \frac{a_{n-1}}{a_{n-2}} \cdots \frac{a_{N+1}}{a_N} \cdot a_N\right| \\ &= \left|\frac{a_n}{a_{n-1}}\right| \cdot \left|\frac{a_{n-1}}{a_{n-2}}\right| \cdots \left|\frac{a_{N+1}}{a_N}\right| \cdot |a_N| \\ &\leq d \cdot d \cdots d \cdot |a_N| \\ &= d^{n-N}\, |a_N| \\ &= d^n \frac{|a_N|}{d^N}.\end{aligned}$$

The series $\sum d^n$ is convergent according to Theorem 2.3.3; since $\frac{|a_N|}{d^N}$ is a constant, this implies that the series $\sum_{n=N}^{\infty} d^n \frac{|a_N|}{d^N}$ is convergent as well. Via Theorem 2.1.4(i) we see that $\sum_{n=N}^{\infty} |a_n|$ is convergent; finally, Theorem 2.1.4(ii) shows that $\sum_{n=1}^{\infty} a_n$ is convergent.

Now consider the case where $C > 1$. Then $|a_{n+1}| > |a_n|$ for sufficiently large values of n; but this means that the sequence $|a_1|, |a_2|, \ldots$ is increasing from a certain term, and therefore $a_n \nrightarrow 0$ as $n \to \infty$. According to the nth term test this implies that $\sum_{n=1}^{\infty} a_n$ is divergent. □

Proof of Theorem 2.2.1: Consider Figure A.4.1. For each $N \in \mathbb{N}$,

$$\sum_{n=2}^{N} f(n) = f(2) + \cdots + f(N) < \int_1^N f(x)dx.$$

If $\int_1^t f(x)dx$ has a finite limit as $t \to \infty$, then the sequence $\sum_{n=2}^N f(n)$, $n \geq 2$, is increasing and bounded, and therefore convergent. It also follows that

$$\sum_{n=2}^{\infty} f(n) < \int_1^{\infty} f(x)dx$$

(it needs an extra argument to see that we have sharp inequality in the limit). So

$$\sum_{n=1}^{\infty} f(n) = f(1) + \sum_{n=2}^{\infty} f(n) < f(1) + \int_1^{\infty} f(x)dx.$$

Similarly,

$$\sum_{n=1}^{N} f(n) = f(1) + f(2) + \cdots + f(N) > \int_1^{N+1} f(x)dx,$$

which implies that

$$\sum_{n=1}^{\infty} f(n) > \int_1^{\infty} f(x)dx.$$

This concludes the proof of (i), and proves at the same time (ii). □

Proof of Theorem 2.4.5: Let us prove the theorem under the (superfluous) assumption that

$$\left|\frac{a_{n+1}}{a_n}\right| \to C \in [0, \infty] \text{ for } n \to \infty. \tag{A.16}$$

First we observe that the series $\sum_{n=0}^{\infty} a_n x^n$ is convergent for $x = 0$; we will therefore assume that $x \neq 0$. Applying the quotient test to the series $\sum_{n=0}^{\infty} a_n x^n$ leads to

$$\left|\frac{a_{n+1}x^{n+1}}{a_n x^n}\right| \to C\ |x| \text{ for } n \to \infty. \tag{A.17}$$

We now split the rest of the proof into three cases. If $C = 0$, we see that

$$\left|\frac{a_{n+1}x^{n+1}}{a_n x^n}\right| \to 0 \text{ for } n \to \infty.$$

According to the quotient test this implies that $\sum_{n=1}^{\infty} a_n x^n$ is convergent; since x was arbitrary, we conclude that the series is convergent for all $x \in \mathbb{R}$.

If $C = \infty$, (A.17) shows that

$$\left|\frac{a_{n+1}x^{n+1}}{a_n x^n}\right| \to \infty \text{ for } n \to \infty;$$

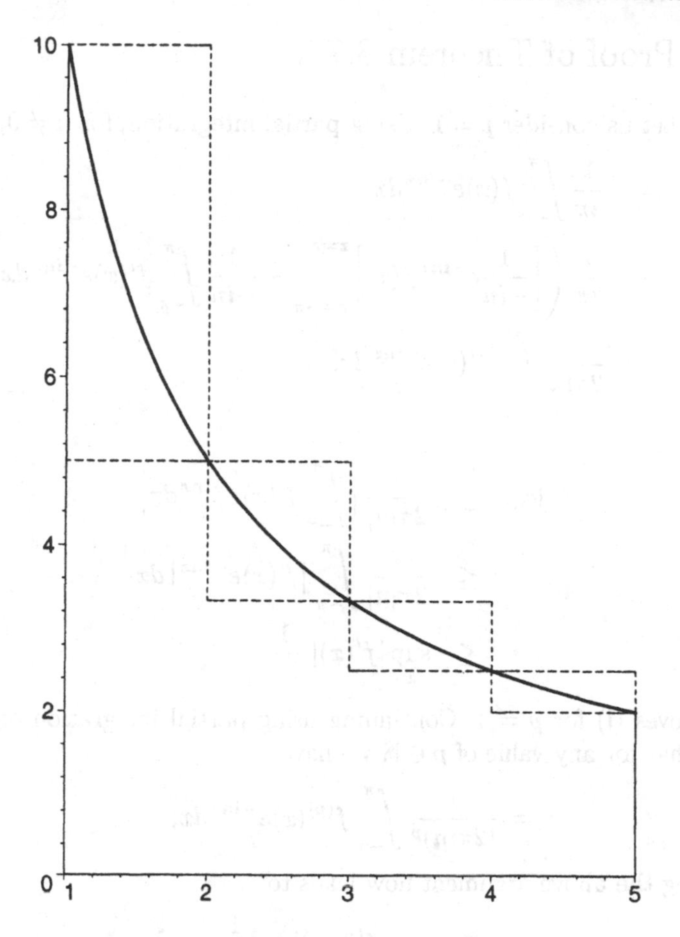

Figure A.4.1 *Graph for a decreasing function $f(x), x \in [1,5]$. We see that $\sum_{n=1}^{4} f(n)$ is an upper sum for the Riemann integral $\int_1^5 f(x)dx$, and that $\sum_{n=2}^{5} f(n)$ is a lower sum.*

in this case the terms in the series $\sum_{n=1}^{\infty} |a_n x^n|$ are increasing, so

$$a_n x^n \nrightarrow 0 \text{ for } n \to \infty.$$

Thus, $\sum_{n=0}^{\infty} a_n x^n$ is divergent for $x \neq 0$ according to the nth term test.

Finally, if $C \in]0, \infty[$, then (A.17) together with the quotient test shows that $\sum_{n=1}^{\infty} a_n x^n$ is convergent if $C|x| < 1$, i.e., for $|x| < \frac{1}{C}$, and divergent if $|x| > \frac{1}{C}$. □

A.5 Proof of Theorem 3.7.2

Proof: Let us consider $p = 1$. Using partial integration, for $n \neq 0$,

$$\begin{aligned} c_n &= \frac{1}{2\pi}\int_{-\pi}^{\pi} f(x)e^{-inx}dx \\ &= \frac{1}{2\pi}\left(\left[\frac{1}{-in}e^{-inx}f(x)\right]_{x=-\pi}^{x=\pi} - \frac{1}{-in}\int_{-\pi}^{\pi} f'(x)e^{-inx}dx\right) \\ &= \frac{1}{2\pi in}\int_{-\pi}^{\pi} f'(x)e^{-inx}dx. \qquad (A.18) \end{aligned}$$

Thus,

$$\begin{aligned} |c_n| &\leq \frac{1}{2\pi|n|}\left|\int_{-\pi}^{\pi} f'(x)e^{-inx}dx\right| \\ &\leq \frac{1}{2\pi|n|}\int_{-\pi}^{\pi} \left|f'(x)e^{-inx}\right| dx \\ &\leq \sup_x |f'(x)|\,\frac{1}{|n|}. \end{aligned}$$

This proves (i) for $p = 1$. Continuing using partial integration on (A.18) shows that for any value of $p \in \mathbb{N}$ we have

$$c_n = \frac{1}{(2\pi in)^p}\int_{-\pi}^{\pi} f^{(p)}(x)e^{-inx}dx;$$

repeating the above argument now leads to

$$|c_n| \leq \sup_x |f^{(p)}(x)|\,\frac{1}{(2\pi)^{p-1}}\frac{1}{|n|^p}.$$

(ii) Regardless of the considered value of $p \in \mathbb{N}$, the assumption (3.34) implies that

$$|c_n| \leq \frac{C}{|n|^3} \text{ for all } n \in \mathbb{Z}\setminus\{0\}.$$

Thus, Theorem 2.6.6 implies that f is differentiable and that

$$f'(x) = \sum_{n\in\mathbb{Z}} c_n in e^{inx}.$$

Iterating the argument shows that f in fact is p times differentiable, and that the announced expression for $f^{(p)}$ holds. □

Appendix B

In this appendix we collect some important power series expansions and Fourier series. For the expansions which appear in the text, we state the relevant equation number.

B.1 Power series

$$\frac{1}{1-x} = \sum_{n=0}^{\infty} x^n, \ |x| < 1 \tag{2.10}$$

$$\frac{x}{(1-x)^2} = \sum_{n=0}^{\infty} nx^n, \ |x| < 1$$

$$e^x = \sum_{n=0}^{\infty} \frac{x^n}{n!}, \ x \in \mathbb{R} \tag{2.15}$$

$$\ln(1+x) = \sum_{n=0}^{\infty} \frac{(-1)^n}{n+1} x^{n+1}, \ |x| < 1$$

$$\sin x = \sum_{n=0}^{\infty} \frac{(-1)^n}{(2n+1)!} x^{2n+1}, \ x \in \mathbb{R} \tag{2.17}$$

$$\cos x = \sum_{n=0}^{\infty} \frac{(-1)^n}{(2n)!} x^{2n}, \ x \in \mathbb{R}$$

$$\arctan x = \sum_{n=0}^{\infty} \frac{(-1)^n}{2n+1} x^{2n+1}, \; |x| < 1$$

$$\sinh x = \sum_{n=0}^{\infty} \frac{1}{(2n+1)!} x^{2n+1}, \; x \in \mathbb{R}$$

$$\cosh x = \sum_{n=0}^{\infty} \frac{1}{(2n)!} x^{2n}, \; x \in \mathbb{R}$$

B.2 Fourier series for 2π-periodic functions

$$f(x) = \begin{cases} -1 \text{ if } x \in [-\pi, 0[, \\ 0 \text{ if } x = 0, \\ 1 \text{ if } x \in]0, \pi[, \end{cases} \quad , \; f \sim \frac{4}{\pi} \sum_{n=1}^{\infty} \frac{1}{2n-1} \sin(2n-1)x \qquad (3.5)$$

$$f(x) = x, \; x \in]-\pi, \pi[: \; f \sim \sum_{n=1}^{\infty} \frac{2}{n} (-1)^{n+1} \sin nx \qquad (3.6)$$

$$f(x) = |x|, \; x \in]-\pi, \pi[: \; f \sim \frac{\pi}{2} - \frac{4}{\pi} \sum_{n=1}^{\infty} \frac{1}{(2n-1)^2} \cos(2n-1)x$$

$$f(x) = x^2, \; x \in]-\pi, \pi[: \; f \sim \frac{\pi^2}{3} - 4 \sum_{n=1}^{\infty} \frac{1}{n^2} (-1)^{n+1} \cos nx$$

$$f(x) = |\sin x|, \; x \in]-\pi, \pi[: \; f \sim \frac{2}{\pi} - \frac{4}{\pi} \sum_{n=1}^{\infty} \frac{\cos 2nx}{(2n-1)(2n+1)}$$

List of Symbols

$\mathbb{R}$:	The real numbers.
$\mathbb{N}$:	The natural numbers: 1,2,3,....
$\mathbb{Z}$:	The integers.
$\mathbb{Q}$:	The rational numbers.
$\mathbb{C}$:	The complex numbers.
i :	The complex unit number.
$\lvert x\rvert$:	Absolute value of complex number x.
$\overline{x}$:	The complex conjugate of $x \in \mathbb{C}$.
$\mathcal{H}$:	Hilbert space.
$L^p(\mathbb{R})$:	For $p \in [1, \infty[$, the space of measurable functions $f : \mathbb{R} \mapsto \mathbb{C}$ for which $\int_{\mathbb{R}} \lvert f(x)\rvert^p dx < \infty$.
$f^{(k)}(x)$:	The kth derivative of the function f.
$C^k(\mathbb{R})$:	The space of k times differentiable functions with a continuous kth derivative.
$\mathcal{F}f = \hat{f}$:	The Fourier transform, for $f \in L^1(\mathbb{R})$ given by $\hat{f}(\gamma) = \int_{\mathbb{R}} f(x) e^{-2\pi i x \gamma} dx$.
χ_A :	The indicator function for a set A, $\chi_A(x) = 1$ if $x \in A$, $\chi_A(x) = 0$ if $x \notin A$.
$\overline{A}$:	The closure of a set A.
supp$f$:	The support of the function f: supp$f = \overline{\{x \in \mathbb{R} : f(x) \neq 0\}}$.
T_a :	The translation operator $(T_a f)(x) = f(x - a)$.
$\langle \cdot, \cdot \rangle$:	The inner product in a Hilbert space.
$\lvert\lvert \cdot \rvert\rvert$:	The norm in a normed vector space.

References

[1] **(D)** J. Berger and C. Nichols: *Brahms at the piano.* Leonardo Music Journal **4** (1994), 23–30.

[2] **(B)** C. M. Brislawn: *Fingerprints go digital.* Notices of the Amer. Math. Soc. **42** no. 11 (1995), 1278–1283.

[3] **(C)** O. Christensen: *An introduction to frames and Riesz bases.* Birkhäuser Boston, 2003.

[4] **(A-B)** O. Christensen and K. Laghrida Christensen: *Linear independence in function spaces.* Normat **51** (2003), 2–14.

[5] **(D)** A. Cohen, I. Daubechies, and J.-C.Feauveau: *Biorthogonal bases of compactly supported wavelets.* Comm. Pure Appl. Math. (1993), 485–560.

[6] **(D)** C. Chui, W. He, and J. Stöckler: *Compactly supported tight and sibling frames with maximum vanishing moments.* Appl. Comp. Harm. Anal. **13** no. 3 (2002), 224–262.

[7] **(B)** J. Conway: *One complex variable.* Springer, 1985.

[8] **(C)** I. Daubechies: *Ten lectures on wavelets.* SIAM, Philadelphia, 1992.

[9] **(D)** I. Daubechies, A. Grossmann, and Y. Meyer: *Painless nonorthogonal expansions.* J. Math. Phys. **27** (1986), 1271–1283.

[10] **(D)** I. Daubechies and B. Han: *The canonical dual of a wavelet frame.* Appl. Comp. Harm. Anal. **12** no. 3 (2002), 269–285.

[11] **(D)** I. Daubechies, B. Han, A. Ron, and Z. Shen: *Framelets: MRA-based constructions of wavelet frames.* Appl. Comp. Harm. Anal. **14** (2003), 1–46.

[12] **(H)** R. J. Duffin and A. C. Schaeffer: *A class of nonharmonic Fourier series.* Trans. Amer. Math. Soc. **72** (1952) 341–366.

[13] **(H)** J. Fourier: *The analytical theory of heat.* Dover, New York. This is a translation of Fourier's original article *Theorie analytique de la chaleur,* publiced in Paris by Didot, 1822.

[14] **(D)** A. Grossmann, J. Morlet, and T. Paul: *Transforms associated to square integrable group representations I.* J. Math. Phys. **26** no. 10 (1985), 2473–2479.

[15] **(C)** K. Gröchenig: *Foundations of time-frequency analysis.* Birkhäuser, Boston, 2000.

[16] **(H)** A. Haar: *Zur Theorie der Orthogonalen Funktionen-Systeme.* Math. Ann. **69** (1910), 331–371.

[17] **(D)** C. Heil, J. Ramanathan, and P. Topiwala: *Linear independence of time-frequency translates.* Proc. Amer. Math. Soc. **124** (1996), 2787–2795.

[18] **(C)** E. Hernandez and G. Weiss: *A first course on wavelets.* CRC Press, Boca Raton, 1996.

[19] **(A-B)** B. B. Hubbard: *The world according to wavelets: The story of a mathematical technique in the making.* AK Peters, Ltd., Wellesley, MA, 1996.

[20] **(C-D)** S. Jaffard and Y. Meyer: *Wavelet methods for pointwise regularity and local oscllations of functions.* Mem. Amer. Math. Soc. 123 no. 587. AMS, Providence, RI, 1996.

[21] **(C)** S. Jaffard, Y. Meyer, and R. Ryan: *Wavelets; Tools for science & technology.* SIAM Philadelphia 2001.

[22] **(B)** A. Jensen and A. la Cour-Harbo: *Ripples in mathematics.* Springer-Verlag Berlin, 2001.

[23] **(B)** E. W. Kamen and B. S. Heck: *Fundamentals of signals and systems using the web and Matlab.* Prentice Hall New Jersey, 2000.

[24] **(D)** P. Linnell: *Von Neumann algebras and linear independence of translates.* Proc. Amer. Math. Soc. **127** no. 11 (1999), 3269–3277.

[25] **(C)** S. Mallat: *A wavelet tour of signal processing.* Academic Press, San Diego, 1999.

[26] **(C)** Y. Meyer: *Wavelets and operators.* Cambridge University Press 1992.

[27] **(B-C)** W. Rudin: *Real and complex analysis.* McGraw Hill, 1966.

[28] **(C)** R. Young: *An introduction to nonharmonic Fourier series.* Academic Press, New York, 1980 (revised first edition 2001).

[29] **(B-C)** M. Vetterli, and J. Kovačević: *Wavelets and subband coding.* Englewood Cliffs, NJ. Prentice-Hall 1995.

[30] **(B-C)** D. Walnut: *An introduction to wavelet analysis.* Birkhäuser, Boston, 2001.

[31] **(B-C)** M. V. Wickerhauser: *Adapted wavelet analysis from theory to software.* A. K. Peters, Ltd., 1993.

Index

adaptive methods, 103
alternating series, 23

Balian–Low theorem, 134
best N-term approximation, 46
biorthogonal wavelets, 130

Cauchy sequence, 64
compact support, 88
comparison test, 19
compression, 46
conditionally convergent, 20
continuous function, 138
continuous signals, 93
convergent series, 16

Daubechies' wavelet, 90
discrete signals, 93
discrete wavelet transform, 94
divergent series, 16
dual, 130

equivalent series, 20
exponential decay, 89

Fourier coefficients, 52
Fourier series, 52
Fourier transform, 77
Fourier's theorem, 58
frame, 130

geometric series, 24
Gibb's phenomena, 54

Haar wavelet, 89
Hilbert space, 64

infinite series, 16
integral test, 22

JPEG, 102

majorant series, 43
Meyer's wavelet, 90
multiresolution analysis, 90
multiscale representation, 118

nonlinear approximation, 74
nth term test, 19

orthonormal basis, 64

Parseval's theorem, 69
partial sum, 16

piecewise differentiable, 57
pixels, 98
polynomial decay, 72
power series, 30
power series representation, 30

quotient test, 21

radius of convergence, 30
reconstruction, 95
Riesz basis, 129
Rolle's theorem, 141

signal, 1, 44
signal transmission, 44
spline, 88
step function, 53
sum function, 34
support, 122

Taylor polynomial, 8
Taylor's theorem, 10
thresholding, 96
tight frame, 130
triangular inequality, 137

unconditional convergent, 20
uniform approximation, 3
uniform continuity, 138
uniform convergence, 42

Wavelet Digest, 92
wavelet packets, 103
wavelet system, 85
wavelet transform, 108
Weierstrass' M-test, 43
Weierstrass' theorem, 5